# AROMATIC ORIGINS

## Chocolate Makers in Cocoa's Heartlands

Alain d'Aboville

Cover design and typesetting by Kristina Konstantinova.
Editing by Karen d'Aboville

alain@chocodabo.com
chocodabo.com

First Printing, 2026
ISBN 978-1-7340040-1-4

# CONTENTS

# PROLOGUE

**In a world** where a chocolate bar can cost more than ten dollars, it's easy to forget that, for most people, chocolate has long been seen as a cheap, sugary commodity. Until recently, it was exclusively mass-produced by large, multinational companies. These companies marketed their products for their uniformly sweet taste and smooth texture. Only a few passionate flavor seekers made small batches of distinctive chocolates, sold in upscale stores and grabbed up by dedicated connoisseurs.

That began to change in 1996, when Robert Steinberg and John Scharffenberger founded the Scharffen Berger Company in Berkeley, California. Their goal was to create a flavorful, complex chocolate that didn't yet exist in the U.S. Their venture helped launch what is now known as the "Bean to Bar" movement. Scharffen Berger emphasized sourcing high-quality cacao beans and managing every step of production with precision in a small laboratory. They even designed and built custom machinery. In 2006, the American confectionery giant Hershey's acquired Scharffen Berger, aiming to gain a foothold in the promising new craft chocolate sector.

During the 2000s and 2010s, the market evolved, and consumers' expectations about what chocolate could be—and what it could cost—began to rise. But while prices for fine chocolate bars increased in the Global North, cacao bean prices in producing countries stagnated, leaving many farmers in an unsustainable situation. By the end of the 2010s, cacao farmers worldwide faced chronically low prices for their beans and sought alternative sources of income. The growing availability of small-scale processing equipment allowed the most entrepreneurial artisan farmers to move up the value chain by making their own chocolate.

Today, a combination of persistent challenges and overall market factors—low yields, limited farmland expansion, shrinking rural workforces, and soaring demand—have driven cacao prices to unprecedented levels. As of now, skyrocketing prices remain above US$7,000–8,000 per ton, and this trend is likely to continue for the next five to eight years.This price surge has caught the attention of investors and governments alike, who now see cacao farming not just as subsistence agriculture but as a promising rural business opportunity.

## The Global Chocolate Context

**Although West Africa** now accounts for over two-thirds of the world's cacao bean harvest, the cacao species (Theobroma cacao) originated in the deep valleys of South America's Amazon basin, where it thrives in a hot and humid rainforest environment. Through early cultural and commercial exchanges, cacao quickly made its way northward and became endemic to Central America. There, it developed unique characteristics and became deeply woven into Mesoamerican civilizations.

In the 16th and 17th centuries, South America's Spanish colonizers introduced cacao to Europe, adapting the original indigenous drink—which was traditionally spicy and bitter—by sweetening it with sugar to suit European palates. Starting in the 17th century, to answer Europe's growing demand, the Spanish crown sent cacao seeds to all its possessions where the climate was adequate. Other European powers soon followed, planting cacao throughout their tropical territories. Then, chocolate in Europe was exclusively consumed as a drink until the second half of the 19th century.

In 1828, the Dutch chemist Coenraad Van Houten built a machine that allowed the separation of the cacao fat (butter) from the cacao mass. The resulting cocoa powder was easier to blend with milk or water than plain ground cacao beans with their full fat content, and the resulting drink was easier to digest. At the same time, the development of steam power made it economically viable to grind and mix large quantities of cacao beans.

Finally, in 1879, the Swiss Rudolf Lindt accidentally created the modern "conche," or mixing machine, that produces a very smooth chocolate texture by thoroughly enrobing the cacao particles with the fat and sugar molecules of the mix. These technical innovations enabled the Fry's company in London to create the world's first modern chocolate bar and ushered the industrial phase of chocolate into the 20th century. By the early 20th century, chocolate had shifted from a handcrafted luxury to an industrial product.

SOURCE: STATISTA.COM USING ICCO DATA

After the First World War, the expansion of European colonies in West Africa and the Far East supplied growing quantities of cacao beans. Following independence, most former colonies developed cacao-friendly policies in order to support rural employment and to secure foreign currencies through export sales. In many countries, such as Côte d'Ivoire—which currently produces 40% of the world's harvest—the growth of cacao farming came at the expense of forests, leaving little available land available to increase harvests. This challenge is compounded by the effects of climate change, which are forcing farmers to relocate some production areas.

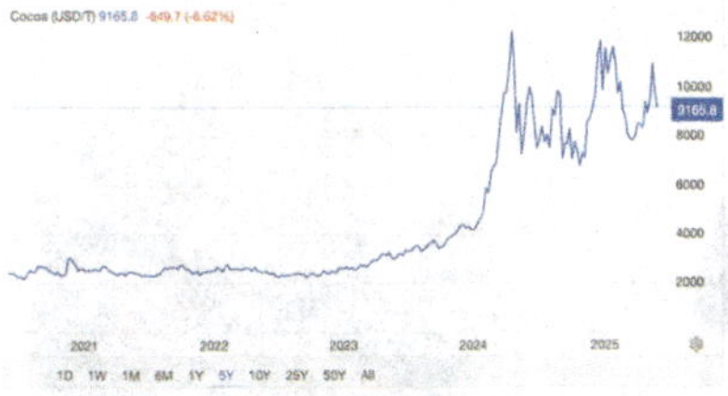


## A Price Shock

**In 2023, five** million tons of dried cacao beans were traded at a value of US$13 billion. Once transformed into chocolate and pastries, they generated a US$130 billion market. That year, the harvest declined by 3% while demand continued to rise, driving cacao prices sharply upward—from US$2,200 to $8,800 per ton, and briefly exceeding $10,000. Since production is unlikely to increase significantly over the next four to six years, cacao prices will probably remain volatile at elevated levels, often above US$7,000 per ton. Cacao beans' current high prices are attracting new entrants, such as Ethiopia, to develop a cacao farming industry. At the same time, many countries like Trinidad and Malaysia, which had previously abandoned cacao due to poor economic returns, are experiencing a revival.

## The Process of Making Chocolate

**Turning cacao beans** into chocolate requires a multitude of steps, each calling on distinct know-how and skills.

### In the field:

## Steps not requiring proximity to cacao fields

**Roasting skills**
CONTROLLED ROASTING BETWEEN 250 AND 320 DEGREES FAHRENHEIT DURING 10 TO 40 MINUTES

**Mechanical skills**
WINNOWING TO REMOVE THE BEANS' SKINS OR HUSKS

**Mechanical and chemical skills**
GRINDING BEANS TO A SIZE PARTICULE BELOW 20 MICRONS

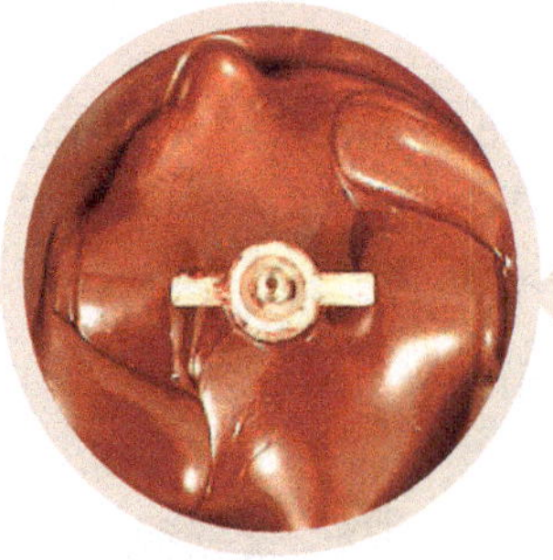

**Mechanical and tasting skills**
CONCHING TO MIX SUGAR AND CACAO CONTENT

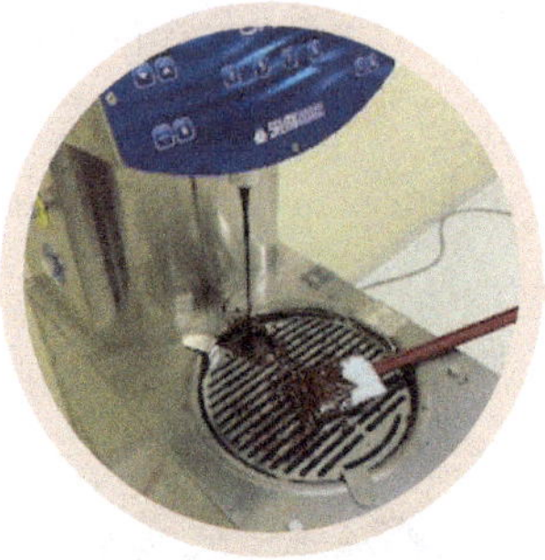

**Mechanical and pastry skills**
TEMPERING TO SMOOTH THE TEXTURE AND SOLIDIFY THE CHOCOLATE WITH A SHINY LOOK

**Mechanical and pastry skills**
MOLDING

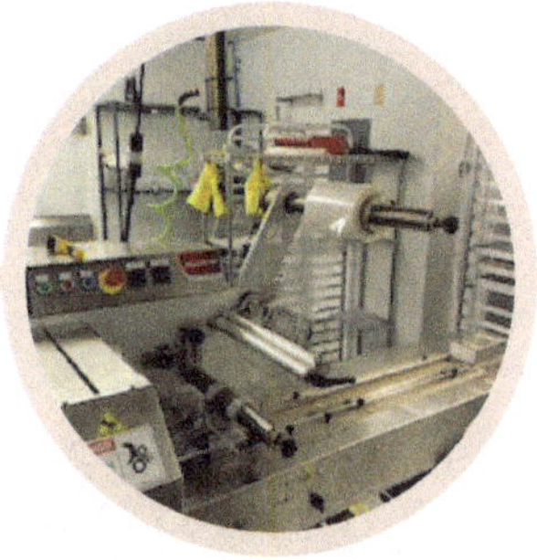

**Mechanical skills**
PACKAGING

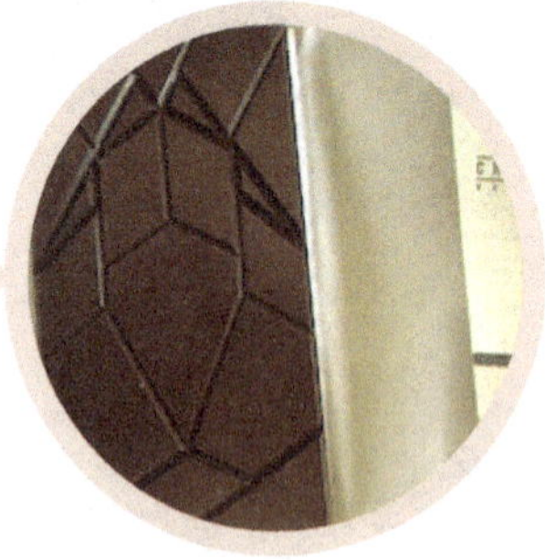

**Commercial skills**
RETAILING

Chocolate aromas are transmitted from the fruit's pulp to its beans during the fermentation process. Flavors can be further enhanced, muted or suppressed at the roasting stage—which is why this stage is often considered one of the most decisive steps in chocolate making.

Producing chocolate also requires a steady supply of electricity. Since energy can sometimes be unreliable or expensive in many cacao-farming regions, most chocolate manufacturing takes place closer to the consuming markets in Europe and the U.S. Logistics add another practical reason: it is far easier and more cost-effective to ship dried cacao beans across oceans than to transport finished chocolate, which is bulky, fragile, and sensitive to heat.

Behind every exceptional bar of chocolate lies a story—of vision, craft, and conviction. The following twelve interviews, followed by a description of a new trend in cacao transportation, bring together some of the most inspiring voices shaping today's chocolate landscape: farmers, makers, innovators, and transport pioneers who tell their stories from tree to bar and beyond. Each of them shares not only their expertise but also their passion for creating a more delicious and responsible chocolate experience. I would like to thank everyone who shared their stories, challenges and expectations with me. It was an amazing learning experience to get to know them not only professionally, but personally as well.

Alain d'Aboville

# BURKINA FASO

CHEF ANDRÉ - ANDRÉ BAYALA

## CHOCOLATE MAKER INTERVIEW

| | |
|---|---|
| Location | OUAGADOUGOU - BURKINA FASO |
| Maker | CHEF ANDRÉ |
| Interviewee | ANDRÉ BAYALA |
| Comments | INTERVIEWED IN HIS BOUTIQUE IN OUAGA DEUX MILLE (BURKINA FASO) |

## OVERVIEW OF BURKINA FASO CACAO / CHOCOLATE SCENE BY THE AUTHOR

**Burkina Faso is** not traditionally a cacao-producing country. The northern half of the country is semi-arid and hot, while only some areas of the southwestern region, bordering Côte d'Ivoire, show potential for cacao cultivation if suitable varieties are identified or created via hybridizing. State-backed programs and private investments have so far failed to create a strong industry. Still, Burkina Faso is closely linked to cacao: countless Burkinabé adults—and many children, some as young as twelve—toil in the plantations just across the Ivorian border. At the time of this interview, Burkina Faso produced only negligible amounts of cacao beans but a growing volume of chocolate, thanks largely to the great efforts of Chef André.

CHEF ANDRÉ BAYALA

## INTERVIEW OF ANDRÉ BAYALA, FOUNDER AND CEO OF CHEF ANDRÉ

### How did you get into chocolate?

**My career path** is undoubtedly unusual because there are very few African chocolatiers, and especially so because I have not benefited from any formal training. Like many young Burkinabés, I had few resources after adolescence, and left for Côte d'Ivoire to look for work in the cooking/catering sector. I started out washing dishes at the university restaurant in Abidjan. It wasn't ideal, but I learned how kitchens and teams function.

Ever since tasting chocolate as a child, I have been fascinated by it—and more broadly by the food industry and what we today call in French *les métiers de bouche* (culinary professions). With Côte d'Ivoire being by far the world's largest producer of cocoa beans, I tried to get closer to the chocolate sector in Abidjan. As you know, many Burkinabés work, often in more than precarious conditions, growing and harvesting cocoa in Côte d'Ivoire. So I gained some firsthand knowledge of cocoa farming there.

In 2014, I met Isabelle Chevalet, a member of the Vaudois Grand Council in Switzerland, during one of her visits to Burkina Faso. Coming from a family of bakers and pastry chefs, I wanted to broaden my palate and deepen my understanding of chocolate. She put me in contact with a master chocolatier from her canton, called Tristan. He welcomed me as an intern in his workshop in a charming village overlooking Lake Geneva and surrounded by vineyards, just above Aubonne. There, I learned how to make chocolate from Bean to Bar. His attention to detail and his passion for flavors were exceptional!

When I returned home to Ouagadougou, he generously gave me some machines that I still use today. They helped me to win the title of "African Champion" of chocolate in 2016. Faced with the lack of competent apprentices, both in chocolate and in bakery-pastry, I set myself the goal of spreading this knowledge to as many people as possible. I created a training center for catering professions in Ouagadougou, where we have since trained nearly a hundred young people. Many immediately find a job after their training with us with restaurateurs, hotels, or other food service providers. The most enterprising have also started their own bakery and pastry shops.

As part of this project, we also set up the largest library of cookbooks in Africa with more than 320,000 books, including over 3,000 dedicated to the subject of cocoa and chocolate. This initiative was made possible by the generosity of Switzerland.

## What is your current situation ?

**Today I have** my main production center in the District of Ouagadougou 2000, another one near the airport, plus two small points of sale in town. I also collaborate with the University of Ouagadougou. Each year, I produce around five tons of chocolate featuring flavors typical of the region, such as ginger, hibiscus, mango or even cashew nuts. Currently, I am experimenting to develop a recipe including locusts and even local caterpillars known as chitoumou. I would like to expand the population of Burkinabé chocolate lovers. While chocolate remains expensive for many Burkinabés, consumption is growing—especially since the COVID-19 pandemic, perhaps because people sought chocolate's relaxing qualities? Since it is very hot here, I temper my chocolate a little longer than in Europe, so that it resists high temperature better, and melts more slowly. I frequently go to Côte d'Ivoire to select and buy the beans I need directly from the farmers, but I am also encouraging growers in southern Burkina Faso to plant cacao trees. This process has started, and I have not given up hope of having good Burkina Faso beans soon.

THE AUTHOR AND CHOCOLATE MAKER TRISTAN IN HIS SHOP IN SWITZERLAND

## Tell me more about your branding?

**I FOCUS MY** branding on my country, Burkina Faso. Although my brand is my name, Chef André, on my packaging I incorporate symbols referring to the history and culture of the country, like its flag or a photo of Thomas Sankara, among others. Sankara was an idealistic, assassinated leader who changed the country's name from Upper Volta to Burkina Faso, meaning "Land of the Upright People" to reflect integrity and self-reliance.

THOMAS SANKARA, REVERED REVOLUTIONARY LEADER OF THE COUNTRY IN THE 1980'S

## What are your main challenges?

**IT IS DIFFICULT** to answer as there are so many. For example, buying beans in the nearby Côte d'Ivoire, south of Bobo-Dioulasso, should be an easy operation by road, but the Ivorian authorities and intermediaries make crossing the border an uncertain and complicated step. Here in Ouagadougou for chocolate production, electricity is both expensive and unreliable. When power cuts unexpectedly interrupt production, not only do the machines obviously stop but also the small air-conditioning units fail. As a result, we often have to restart batches from scratch. At the staff level, I need to train young people not only in the professions of chocolatier or pastry, but also in essential business behaviors, since schools no longer provide the necessary foundations for professional life. Packing is also problematic: local options are expensive and of inconsistent quality. Because I don't have a large volume and also due to currency restrictions, I cannot buy packaging from China. So rather than pointing to a single obstacle, I see all these issues as the daily realities to be managed. It's tough, but that's life — I don't complain.

## How do you see your business in 5 years?

**I BELIEVE THAT** local and regional chocolate consumption will grow. I hope to open a production center and a shop in Bobo-Dioulasso, and for local farmers in the southwest to cultivate cacao so that I can produce 100% Burkinabé chocolate. My vision is not only to avoid importing cacao beans but also to "move up" in the chocolate value chain by exporting semi-finished products such as nibs (fermented, roasted and broken beans) and eventually cocoa liquor. It frustrates me that West Africa, which produces more than three-quarters of the world's cocoa, only retains around 5% of the value!

## What is a good chocolate?

**I am very** open to what people like in a chocolate product. Good chocolate is what people enjoy eating. I make chocolates with a multitude of ingredients because I think the mix of flavors, scents, and sensations can be pleasant. Of course, the chocolate must be well made, with a smooth texture and balanced ingredients, but what matters is the pleasure it gives in the mouth. High cacao percentages, added cocoa butter, emulsifiers, inclusions, or unusual combinations—all are welcome if the sensation while tasting is satisfying, agreeable, and surprising.

THE 70% GINGER-FLAVORED BESTSELLING BAR

## How do you see the chocolate industry in 10 years?

**Faced with cacao** prices inevitably rising, large international chocolate companies will likely adapt by launching new products using less cacao and more flavor substitutes. This shift will allow quality chocolatiers to stand out more clearly and thus justify their prices.

# UGANDA

## LATITUDE CHOCOLATE - JEFF STEINBERG

## CHOCOLATE MAKER INTERVIEW

| | |
|---|---|
| LOCATION | KAMPALA - UGANDA |
| MAKER | LATITUDE CHOCOLATE |
| INTERVIEWEE | JEFF STEINBERG |
| COMMENTS | REALIZED AT LATITUDE COFFEE SHOP IN KAMPALA |

## OVERVIEW OF UGANDA CACAO / CHOCOLATE SCENE BY THE AUTHOR

**CACAO WAS INTRODUCED** to Uganda as early as 1901 by the British and first exported in 1917. However, the volatile and large fluctuations of the price of the beans soon discouraged London investors, who withdrew financing, and cacao farming was all but abandoned within a few years. In 1958, the colonial government, eager to diversify Uganda's agricultural exports, attempted to re-introduce cacao cultivation by financing new farms. This effort was successful, and the cacao bean harvest peaked in 1960, becoming the country's fourth-largest food export after coffee, tea, and fish.

State support declined during the early years following the country's independence from Great Britain. During the Idi Amin Dada dictatorship the business environment became hostile to large-scale investment. It was only in the 1990s, with the liberalization of farming policies, that cacao production began to grow again.

Today in Uganda, it is estimated that 20,000 hectares are dedicated to cacao farming, primarily in the Western and Central districts. Around 15,000 farmers cultivate and harvest roughly 37,000 tons of beans annually (2023).

The varieties grown are mostly Trinatarios, including an aromatic sub-variety of Forastero from the upper Amazon called Amelonado, along with several new hybrids—all recognized as "fine aroma" cacao capable of fetching premium prices.

JEFF STEINBERG IN HIS PRODUCTION SITE

## INTERVIEW OF JEFF STEINBERG, FOUNDER AND CEO OF LATITUDE CHOCOLATE

### How did you come to chocolate?

**I WENT TO** Sierra Leone in 2011 to work with small cacao farmers as a social worker for the Gates Foundation. I was fresh out of my B.A. in economics from Cornell. I left the Gates Foundation in 2016 to work on a cotton plantation in northern Uganda, near South Sudan. It was then that I realized that cacao had far greater potential for positive social improvement. So, in 2017 I launched my cacao journey by buying a 25-ton container of cacao beans for export—effectively becoming a cacao broker. From the start and of course now, my vision and our mission is to be a social enterprise. That is why we are a Certified B Corporation, recognized for meeting the highest standards of social and environmental behavior and responsibility.

For us, cacao is a tool for social improvement—a way to create financially rewarding local jobs that improve the lives of communities around Uganda. As the craft or specialty chocolate segment was growing steadily in the U.S. and Europe, thus opening a market for our premium Uganda beans, we grew into a significant Ugandan cacao broker. Alongside exporting beans, we also began processing them here in Uganda in order to also sell cocoa powder and cacao butter. I like to say that we are a Ugandan crew of radical and cocoa-obsessed farmers, agronomists, roasters, and chocolate makers dedicated to using chocolate as a force for good in our community. Today we source beans from over 2,500 certified organic farmers and transform them into award-winning chocolate right here at their origin.

THE AUTHOR WITH JEFF STEINBERG AND TEAM

## What is the present situation of Latitude Chocolate?

IN TERMS OF volume, we currently process about 20 tons per year. We sell beans to the specialty market in the U.S., Canada, U.K., the E.U., and Japan. I estimate that there are about 600 potential buyers of premium beans in North America, and substantially fewer in Europe. Besides the brokerage activity, we opened our own Bean to Bar café, with its own chocolate factory, here in the center of Kampala. This allows us to showcase the high aromatic quality of Ugandan beans, mostly coming from the Bundibugyo region, and to serve a small but growing local market. Our team of Ugandan chocolate makers as well as pastry chefs produce the whole range of chocolate products, from traditional chocolate bars and couverture chocolate to sophisticated bonbons, ganaches, pastries, and inclusions. All the ingredients are sourced locally. Currently, about 80% of our revenue comes from trading, but our long-term goal is to balance this with chocolate production, with a 50/50 split.

## What are your main challenges?

WE HAVE THE same issues as many chocolate businesses everywhere. For us, it is human resources, bean sourcing, and equipment. In Kampala, there is a large pool of young workers, but training and retention are often difficult. Young employees tend to move on quickly, which means high turnover. Uganda has good cacao bean varieties, and we operate our own fermenting center in order to ensure consistent fermentation results. But geography sometimes works against us: farmers in remote areas must deliver fresh beans within 12 hours of harvest, which can be logistically complex. As for equipment, being in a small, landlocked country makes it hard to access machinery or spare parts. Fortunately, local mechanics are efficiently resourceful, so this is a manageable issue.

## How do you see your activity in 5 years?

I EXPECT OUR retail business to grow substantially. At the same time, the trading/brokering operations may expand to more cacao regions inside Uganda and perhaps even to neighboring countries.

## What is a good chocolate?

THERE ARE NOW excellent beans across Africa being transformed to make delicious chocolate. For me, good chocolate is where the bean shines through—when you stop blending millions of beans and instead let individual origins speak, you start tasting what makes each bean unique. However, goodness is not just about chocolate tastes but more importantly about the ethics of how the chocolate is made. We emphasize working closely with cacao farmers and pay premium prices for beans that are sustainably grown, which contribute to the creation of jobs and increase farmers' share of the profits.

## How do you see the cacao/chocolate business in 10 years?

ON THE LOCAL level, I think cocoa farming in Uganda will continue expanding, especially if, as I suspect, cacao prices keep rising. Globally, there is no reason for the expansion of chocolate consumption to slow. We always hear of the Chinese and Indians becoming chocolate eaters and their markets are only just beginning to grow. Higher prices will attract new and sustainable cacao producers. For chocolate to become truly sustainable, chocolate makers must take into account the importance farmers play in the cacao value chain. I foresee a substantial increase in the number of Farm to Bar chocolate makers, as rising bean prices and tighter supply and scarcity due to disease or increased demand will cause chocolate makers to secure their sources of beans and ensure their quality directly. Just as major brands like Valrhona and other big chocolate companies have invested in plantations, more companies will own farms to guarantee supply and quality. So the future of chocolate, in my view means more cacao, more sustainability, and inevitably, more expensive chocolate.

# BELIZE

CORRIDGEREE - RUTH MOLONEY

## CHOCOLATE MAKER INTERVIEW

| | |
|---|---|
| LOCATION | HOPKINS - BELIZE |
| MAKER | CORRIDGEREE |
| INTERVIEWEE | RUTH MOLONEY |
| COMMENTS | REALIZED OVER TWO DAYS AT THE PLANTATION |

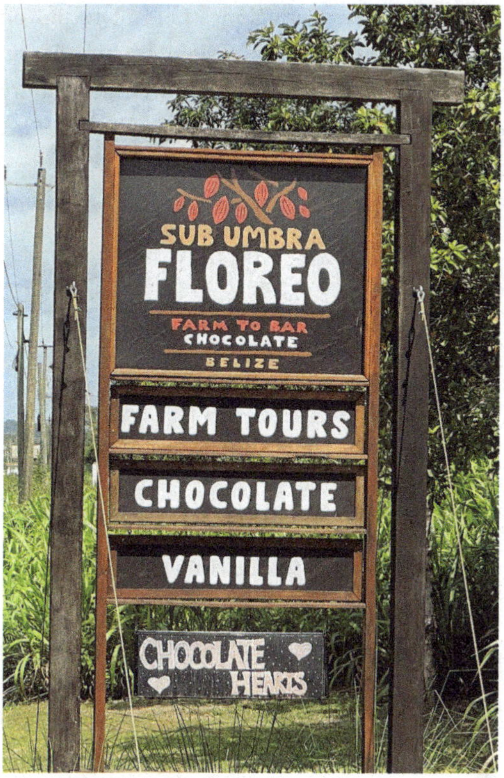


CORRIDGEREE'S MOTTO:
"UNDER THE SHADE IT FLOURISHES"

## OVERVIEW OF BELIZE CACAO / CHOCOLATE SCENE BY THE AUTHOR

**Belize has a** deep-rooted history with growing cacao. The Olmecs—living in what is now southern Mexico—and later the Maya have been cultivating the plant for thousands of years in the isthmus region of Central America. In fact, the word chocolate is derived from the Maya word *xocolatl*. Although the exact circumstances remain unclear, many believe that Belize is the place where Christopher Columbus first encountered cacao beans in 1520. At that time, beans were simply dried (not fermented), ground and mixed with hot water. The resulting drink was fatty, frothy and rather bitter, often flavored with multiple spices such as chili, cinnamon, and others.

The Spanish soon developed cacao plantations, but production in Belize was quickly outpaced by operations in the Mexican colonies and Caribbean islands, particularly in Trinidad and Jamaica. During British colonization and years between the two world wars, cacao farming stagnated. By then, large European chocolate companies had secured more economical supplies from Brazil, Equatorial Guinea and Sao Tome, and eventually West Africa.

Efforts to revive cacao cultivation in Belize began in the 1970s with the introduction of aromatic seeds from Trinidad (called Trinatario), Costa Rica, and Ecuador. However, cacao struggled to compete against more profitable crops like sugar cane and palm oil, and these attempts fizzled. Cacao plantations dwindled, and trees became wild, diseased, and largely nonproductive.

In the 1980s, the American company Hershey established operations in southern Belize but abandoned them when cacao prices collapsed in the early 1990s. A turning point came in 1994 when Green & Black's, then a young British organic foods company (now owned by the American giant Mondelez), launched its Maya Gold range of organic and fair-trade chocolate bars, sourcing beans from the Toledo district, close to Guatemala, but manufactured in the U.K. Supported by Cadbury and Green & Black's, an NGO called the Toledo Cacao Growers Association (TCGA) was created in 1990—today with more than 1,300 members, many from indigenous Mayan villages—to produce quality organic beans.

With a high percentage and heritage of ancestral genetic varieties, Belize cacao is generally very appreciated and enjoys an image of high aromatic quality. In 2023, the country's harvest totaled about 200 tons, divided mainly between Maya Mountain Cacao and the TCGA. Small specialized operations have recently been launched, both in the south (Finca Chocolat) and on the coast (Corridgeree) but are still ramping up production. Belize now has over a dozen artisanal Bean to Bar chocolate makers, all small and most exclusively selling in the country. Several are indigenous Mayan-owned businesses, such as Ixcacao. Others include Cotton Tree Chocolate, Floreo Chocolate, AJAW Chocolate, Cheil Mayan Chocolate, Goss Chocolate, Belize Chocolate Company, Lamanai Chocolate, Dragon Chocolate, La Especial Chocolate, Moho Chocolate Company, and Copal Tree Chocolate.

RUTH MOLONEY, FOUNDER OF THE CORRIDGEREE VENTURE

## INTERVIEW OF RUTH MOLONEY, FOUNDER AND CEO OF CORRIDGEREE

### How did you come to chocolate?

I WAS BORN in Australia and grew up in Queensland, near the Great Barrier Reef. I have always been very interested in everything agricultural. My career began as a sugar trader for CSR Limited (Colonial Sugar Refining), in Australia, where I worked for 13 years, before moving on to Nestlé. At Nestlé, I handled complex financial operations like hedging. Life at such a massive company—260,000-employees worldwide, could have been very cushy, but I eventually felt the need for a change. In 2010, I joined the British commodity investment firm Armajaro based at their New York office. Trading had become increasingly automated, with quantitative models replacing human decision-making. I quickly realized that the cacao market is dominated by just four giants: Olam, Cargill, Ecom, and Barry Callebaut. Their enormous, combined pressure on West Africa, which produces three-quarters of the world's cacao, seemed unsustainable. Because of the effect of climate change, yields in West Africa are bound to decline, especially through increased diseases. Then, in 2014, Armajaro was bought by Ecom, and I was laid off—at the same time that my future business partner, Tino Cota, left Barry Callebaut. After taking a few months to cycle across the U.S., we started doing consulting work. One of these projects involved a coconut plantation in Belize, which is how I came to this country. That's when I discovered Belize's century-old cacao agroforestry tradition, its rich genetic diversity of cacao and vanilla, its proximity to the U.S. and its relatively good infrastructure. Together, these factors convinced me that Belize was an ideal spot to launch an agroforestry business to develop a sustainable cacao and vanilla farm. In 2016, Tino and I teamed up, and we started the process of buying land in Belize near Silk Grass, naming the project Corridgeree. We now have 75 acres (30 hectares) of land and will leave half of it untouched in order to protect the native vanilla species growing there. We started planting our first vanilla seedlings in 2019, at a time when the world price for vanilla was around US$600 per kilo, whereas it averages around US$80 today.

## Where is the name Corridgeree coming from?

**Corridgeree is named** after a long-established family farm in Australia. The name comes from a local Aboriginal dialect and means "Round Hill." Our chocolate brand, *Sub Umbra Floreo* ("Under the shade I flourish"), takes its name from the Belizean national motto, which appears beneath a mahogany tree on the country's coat of arms and national flag. The motto is a well-known reference to Belize's forests and its history as a former British colony (Author's note: Belize gained independence in 1981 and remains part of the British Commonwealth). For our business, it symbolizes how we cultivate under shade trees and use methods that help protect the forests.

## What is the present situation of Corridgeree?

**In terms of** installation and land preparation, we started by preparing the terrain. We then implemented a drip irrigation system and planted the shade trees. We also built a complete organic fertilizer system using worms and coconut husk waste, which produces an amazingly efficient and chemical-free compost. Our cacao nursery is already producing seedlings that will eventually be grown under shade trees. We expect cacao to represent about 10% of our turnover in the long run, with vanilla and chocolate sales—including cacao-tourism and farm tours—making up the rest. Regarding vanilla, we have now planted about 60% of our target number of trees and are on course to reach 100% by 2026. Our first small harvest is expected this year. For this batch, we already have a contract with a U.S. vanilla extractor. All over Belize, cacao suffered an exceptionally serious drought last year, and everywhere the harvest has been reduced both in volume and in quality as both the pods and the inside beans were smaller than usual. Despite this, we are using our fermentation center to produce quality beans. Still, vanilla remains our top priority, with a target of 1 to 2 tons per year, provided we secure reliable buyers.

## What are your main challenges?

**Our current challenge** is to complete our startup phase and to fully implement a formal structure and organize the various activities of the business. This is realized by people, so our paramount issue is human resources. We employ 20 local staff, but competing with the attractive opportunities provided by tourism and service sectors is difficult—those jobs are less physically demanding than agriculture, which involves heavy labor in the heat and sun, bending, and getting dirty, and so forth. We are structuring the business, writing clear procedures, and building teams around specific tasks. For example, we maintain an up-to-date, detailed inventory of everything, not only equipment but also trees, seeds, machines, small tools, etc. We maintain a data-driven dashboard to monitor our activities, output, and main costs like energy and water.

THE AUTHOR IN THE CHOCOLATE LAB WITH CORRIDGEREE TEAM

VANILLA ON HANGING POLES UNDER SHADING NETS

## How do you see your activity in 5 years?

**We started this** venture with a master plan. However, the on-the-ground realities of actually "doing things" always force you to adapt to the unavoidable climatic evolutions and their impacts on farming as well as to respond to changing market demands. Having said that, my goal in 5 years is to have a sufficiently well-organized and structured business that I can hand over to managers. By then, we should have an ongoing production and post-harvest treatment for the vanilla (the whole chain from flowering to packaging, sweating, and drying) with an agile sales system allowing us to ride the inevitable roller coaster of the market. Our chocolate should by then be a recognized brand in Belize, with a sustained level of sales nationally in multiple outlets including airports, resorts, and hotels. Note that we do not plan on exporting chocolate, except for rare, unique opportunities like special corporate or family events. All this should be generating enough revenues to produce a respectable profit. In other words, the business should be self-sustaining, and this would make it attractive to investors and competitors alike.

REGENERATIVE AGRICULTURE TAKES TIME

CORRIDGEREE GROUNDS

## What is a good chocolate?

**For me, a** good chocolate is a total sensory experience, with the aromas as the centerpiece. Whether fruity, chocolatey, floral, or others, the chocolate must develop significant flavors when smelling it. Texture matters, too: it should progressively melt slowly in order to prolong these aromatic sensations into a mellowing bouquet in order to finish the journey with a long and pleasant aftertaste. Beyond this theory, there inevitably is a subjective component to chocolate tasting.

## How do you see the cacao/ chocolate business in 10 years?

**Ten years feels** like an eternity today. I think the current price spike *(Author's note: cacao exchanges set the price close to US$7,000 per ton, 130% peak from last year)* won't significantly benefit farmers globally, since prices are heavily hedged and profits absorbed along the supply chain before reaching farmers. West Africa's declining output won't be readily compensated by newcomers in Asia, Central America, or Africa, meaning that the price hike is likely to become a long-term trend. As for the demand, there is no reason to believe it will drop; if anything, it will keep on increasing. The question is whether chocolate will experience a kind of "Starbucks effect" on as we saw on coffee? By that, I mean a simultaneous increase in quality and sales price worldwide, rather than a reduction in quality to fight price increases. Despite the recent examples of coffee, beer, or pasta that point to a simultaneous increase in price and quality, I cannot say. My hope is that quality wins.

# COSTA RICA

## IGUANA CHOCOLATE - VICTORIA CHANDLER & JORGE GARCIA

## CHOCOLATE MAKER INTERVIEW

| | |
|---|---|
| Location | COSTA RICA |
| Maker | IGUANA CHOCOLATE |
| Interviewee | VICTORIA CHANDLER & JORGE GARCIA |
| Comments | INTERVIEW ONSITE, AT THE PLANTATION ON THE FOOTHILLS OF LA CANGREJA NATIONAL PARK, NORTH OF THE PUNTARENA PROVINCE CLOSE TO THE PACIFIC COAST. |

## OVERVIEW OF COSTA RICA CACAO / CHOCOLATE SCENE BY THE AUTHOR

ONE OF COSTA RICA'S AROMATIC VARIETIES

**Like much of** Mesoamerica, cocoa farming in Costa Rica dates back to pre-Columbian times. Indigenous people cultivated cacao for both culinary and ceremonial uses, creating the frothy beverage xocolatl, flavored with spices, and consumed during rituals and social gatherings.

In the early 16th century, cacao farming expanded to meet the growing demand from the Spanish crown. Plantations were concentrated in the Caribbean lowlands, where climate and soil conditions were favorable. By the 18th and 19th centuries, cacao had become a significant export crop for Costa Rica. After independence from Spain in 1821, the government promoted its cultivation to diversify the economy beyond coffee and sugar and to support rural development. Annual harvests then reached roughly around 2,000 tons.

During the late 19th century, cacao diseases like witches' broom and frosty pod rot devastated plantations. The simultaneous rise of coffee farming further diminished cacao's importance. Nevertheless, small-scale farmers continued cultivating cacao, often integrating it into agroforestry systems alongside bananas and other crops.

In the 20th century, Costa Rica cacao farming shifted towards sustainable practices and organic methods. Farmers embraced agro-ecological techniques, emphasizing biodiversity, soil health, and organic cultivation, with harvest volumes remaining around 2,000 tons.

In the 1990s, a dramatic wave of another fast-spreading disease (Monilia) and the sky-rocketing demand for palm oil pushed farmers away from cacao. Today, despite Costa Rica's reputation for high-quality beans and its fine aroma status, the country produces less than 1,000 tons of dried beans per year. Most farms are located on the Caribbean coast (Talamanca and Huetar Norte) with a handful of "new farmers" on the Pacific side.

## INTERVIEW OF VICTORIA AND JORGE, FOUNDERS AND CEOS OF IGUANA CHOCOLATE

### How did you come to chocolate?

*Jorge:* **My grandfather** had a 15-hectare farm here. In 1985, when cacao farming was decimated by the Monilia fungus, CATIE (Costa Rica's Tropical Agricultural Research and Education Center) started a program to identify the most disease-resistant cacao varieties for each region and terroir. We joined a newly created cooperative covering 200 hectares. My grandfather planted different seeds across his property so that CATIE could evaluate which hybrids were most disease-resistant. Due to the extremely low cacao prices at the time (and for decades) and the devastating effect of the Monilia, many farmers—including my family—switched to palm (for oil), coffee, bananas, or other cash crops. My family as well became disinterested in cacao farming. Cacao trees remained unattended but continued growing on their own. Years later, we revitalized our regenerative farming activities using some of this rootstock in different planting areas. But our cultivated cacao area decreased from 15 hectares to 8, and we now farm about 4 hectares in agroforestry systems.

*Victoria:* **I am** originally from the U.K. and came to Costa Rica around 20 years ago as a volunteer for an NGO promoting and developing sustainable habitats. Jorge was among the people involved in the projects I worked on, and we eventually married. I then concentrated my efforts on the cacao farm. At that time, the Bean to Bar movement had not happened yet, and making chocolate in the middle of the jungle seemed a rather risky endeavor.

LA IGUANA LOGO

THE SHOP

At that time, we worked with volunteers from all around the world to implement our model of organic, bio-diverse agriculture. In 2006, we experimented with making "rustic" chocolate at the farm. By roasting beans in a pan over a wood fire, then hand-peeling them and grinding them with a pestle, we were able to produce small quantities of disks or balls of grainy chocolate. It was rewarding but neither particularly tasty nor economical. So we started searching books and the web for any piece of advice for or examples of making truly amazing artisanal chocolate. With guidance from Clay Gordon's book *Discover Chocolate*, John Nanci's now-well-known website *ChocolateAlchemy.com*, and the support of our short-term volunteers, in 2015, we decided to acquire a "real" 11 kg Santha stone grinder, which marked our entry into the world of professional chocolate makers. But not everything went smoothly. Ninety-nine percent of Costa Rica's electricity is from renewable sources, but supply remains both unreliable and expensive. We spend about US$400 per month to keep our five small "Premier" grinders turning and our small chocolate laboratory cooled. By the late 2010s, there were many Bean to Bar chocolate makers in Europe, and during our travels to the U.K. *(note: where Victoria has family)*, we visited many of them to learn from their practical experience. Chocolate shows also provided valuable knowledge.

ANOTHER VARIETY

## Why the name "Iguana Chocolate" ?

**In the late** 2010s, volunteers were still working at the farm for extended periods of time. We were looking for a name for the burgeoning chocolate business. Something that would work in multiple languages, including local ones. We couldn't agree on anything meeting these criteria. During a brainstorming session in the field, a Mexican/Canadian volunteer saw a small iguana wandering around our feet. That gave him the idea of the name "Iguana Chocolate," which immediately resonated with everyone.

## What is the present situation of "Iguana Chocolate"?

**We are a** true Farm to Bar chocolate maker. This is an important distinction because it means every ingredient included in our chocolate is farmed by us. We grow vanilla and all the fruits we incorporate in our range of flavored chocolate bars, from mango to hibiscus to kumquat, etc. We sell our bars at the shop here at the plantation along with a whole array of related products such as herbal medicines, bags, books, honey, soap, candles, mugs, paintings, all locally produced. We also distribute worldwide through *laiguanachocolate.com* through a drop shipper partner. Chocolate tours have become a key part of our business. Our chocolate business has recently become profitable, which allows us to plan further expansion.

We now exclusively employ local professional farmers and no longer rely on volunteers. We want to expand our product sources. The Pacific region is, by far, not the main cacao region of Costa Rica. We would like to incorporate raw materials from other regions in our products and also retail products from other parts of the country.

## What are your main challenges?

**Operating in a** remote, hot, and humid climate makes daily tasks more difficult. It is hard to identify or rank the severity of each of the issues we face every day. Beyond the permanent logistical headaches, extreme weather variability is our main concern. For example, at this time of year, the soil should be fairly dry and dusty, and the pods growing happily towards the main harvest season around May, prior to the rainy season. But this year it has been raining nearly every day, the landscape is extremely green, the soil and air are humid, and the pods do not grow much in size.

TOTAL INTEGRATION WITH NATURE

ASSESSING FERMENTATION LEVELS

### How do you see your activity in 5 years?

**What we want** to see in 5 years is a diversified business incorporating multiple activities. Iguana Chocolate will produce its own chocolate but be making chocolate and chocolate products with beans from the best sources in Costa Rica as well, including other Bean to Bar makers. We want to build a regular stream of repeat local customers. People who will come to our plantation and shop as part of their routine to buy the best chocolate product in the country. At the same time, we want to increase Iguana Chocolate's tourism side and chocolate-related tourism in general. With the growing influx of tourists, both national and international, we will organize more courses and teaching events. Finally, we would like to host chocolate "meet-ups" to share information, knowledge, and experiences among like-minded Bean to Bar makers world-wide.

### What is a good chocolate?

**A good chocolate** must be "the whole combo". Meaning it must tick each and every box of the chocolate-making process. Quality beans must be organically farmed and ethically harvested. If not from regenerative systems, they should, at least, come from agroforestry farms. As far as tasting goes, the texture doesn't matter much to me, but the flavors are key and must be numerous. I want a crescendo flavor journey while the chocolate is melting in my mouth. Personally, I do not like acidity and enjoy most the caramel notes and red fruit tones.

### How do you see the cacao/chocolate business in 10 years?

**By continuing the** current trend, I expect many more Bean to Bar makers will emerge in the country of the beans' origin, like ourselves. This will be in response to a growing local demand and in order to export to the "traditional" markets (Europe and the U.S.). Globally I believe the industrial chocolate companies will likely offer more flavorful, high-quality products. Seeing how many Bean to Bar makers have multiplied in Europe and the U.S., there will likely be a consolidation period where the most successful craft chocolatiers will be bought by larger companies, while others may disappear. Overall, due to media pressure and multiple chocolate shows, chocolate consumers will become more educated, understanding the whole chocolate process and valuing ethically produced, high-quality chocolate. They will be willing to pay a premium to support farmers and ensure sustainable practices.

VICTORIA IN FRONT OF HER FERMENTATION BOX

# ECUADOR

VILLAKUYAYA - TANIA MOLINA

## CHOCOLATE MAKER INTERVIEW

| | |
|---|---|
| **Location** | ECUADOR |
| **Maker** | VILLAKUYAYA |
| **Interviewee** | TANIA MOLINA |
| **Comments** | REALIZED AT WASHINGTON D.C. CHOCOLATE SHOW<br>FOLLOWED BY VARIOUS ZOOM MEETINGS |

TANIA MOLINA AT THE WASHINGTON D.C. CHOCOLATE SHOW

## OVERVIEW OF ECUADOR CACAO / CHOCOLATE SCENE BY THE AUTHOR

**Ecuador is considered** the cradle of cacao. Cacao trees first appeared thousands of years ago in the upper Amazon basin along its main rivers. Thanks to the remoteness of some of these valleys, cacao has been able to grow without significant genetic interference, giving rise to unique varieties such as the Arriba and Nacional. Commercial plantations already existed before the arrival of the Spanish colonizers. With its ideal soil and climate, Ecuador offered perfect conditions for cacao farming. The Spanish expanded cacao cultivation to meet Europe's increasing demand for cacao beans.

Until the early 20th century, Ecuador ranked among the world's top five cacao producers. However, with the exponential growth of African cacao plantations, Ecuador's position declined. Today, the country produces close to 400,000 tons annually, representing about 8% of global output.

In the late 1960s, a devastating cacao disease prompted Ecuadorian agronomists to develop a resilient and highly productive hybrid: *CCN-51* (Coleccion Castro Naranjal). Due to its high yields, this variety spread quickly and now accounts for nearly two-thirds of Ecuador's production. Unfortunately, *CCN-51* lacks the refined aromatic profile needed to qualify as fine aroma cacao. Still, Ecuador's reputation for flavorful beans endures and is even enhanced by the fact that the world's most expensive chocolate bar, *To'ak* (*www.toakchocolate.com*), is made from Ecuadorian beans.

Ecuador's dynamism is also reflected in its premium chocolate industry. Two notable makers, Pacari (founded by Santiago Peralta and Carla Barbotó, *www.pacari.com*) and República del Cacao (*www.republicadelcacao.com*), produce locally and sell internationally. The latter is majority-owned by the French group Saviencia (*www.savencia.com*), which also controls French chocolate specialty maker Valrhona.

## INTERVIEW WITH TANIA MOLINA, FOUNDER OF VILLAKUYAYA

### How did you come to chocolate?

I WAS BORN in Ecuador and lived there until my late twenties. I worked as an architect in Quito for several years. In the early 2010s, my brother was accepted to New York University, and in 2014 my parents encouraged me to go visit him in New York. Once there, I decided to get involved in chocolate, one of Ecuador's main products, drawing inspiration from and mimicking my grandmother's way of working with it. I visited every chocolate shop I could find, contacted the Ecuadorian Trade Office, attended the Fancy Food Show, and spoke with as many chocolate professionals as I could.

Back in Quito, I travelled to top cacao-growing regions such as Esmeraldas, and purchased beans. I rented a professional chocolate laboratory in Quito and produced over 2,000 chocolate bars that I brought back to the U.S. in 2015. I received a warm welcome from stores, most of them high-end food shops in Manhattan, when asking them to carry my bars. Yet by Christmas, they had sold none.

The following spring, I returned to the Fancy Food Show again with my bars. I also sought advice from the Fine Chocolate Industry Association (FCIA) and met Clay Gordon, founder of the online community The Chocolate Life. Having worked in cacao/chocolate since the mid-2000s, he is a pillar of the Bean to Bar movement, with an extensive understanding of the product and the market. Following his guidance, I redesigned my packaging to a more elaborate one—a slim white box with a stylish cacao fruit charcoal drawing on the front, and detailed technical and storytelling elements on the back. The new look projected knowledge, style, and high quality. Almost immediately, New Yorkers began not only buying my bars but also returning for more.

OUR BESTSELLER IN 2024

### Where is the name "Villakuyaya" coming from ?

MY LATE GRANDMOTHER'S philosophy greatly influenced me. She was a sort of shaman in our village, always experimenting with herbs and natural ingredients. Her relationship with these plants was quasi-mystical. In Quechua, our indigenous language, *kuyaya* means grandmother. Grandmother's behavior around natural ingredients like cocoa beans was a ritual. I envisioned my chocolate project as a home for her spirit—a *villa* where she could live on. That is how *villakuyaya* was born.

BEING "IN THE FIELD" IS ESSENTIAL

**What is your present situation?**

**Today I run** a well-balanced operation spread between Ecuador, where production takes place, and the U.S., where most sales occur. I also practice yoga, which is an important part of my identity. Each spring, I travel to Esmeraldas and make a point of personally visiting the plantations and meeting the farmers, ensuring integrity in sourcing while avoiding corruption and supply risks. Beans are delivered to the chocolate laboratory in Quito during the summer, where the city's cool, dry climate provides ideal processing conditions. Being perched at 9,350 feet (2,850 meters), Quito's weather is never hot or humid like where the cacao plantations are in the northern and western regions of the country. Production is then air-shipped to New York in the early autumn.

Beyond the U.S., I sell internationally thanks to many participations in international chocolate festivals and trade shows. I attended the *Chocoa* festival in Amsterdam three years in a row (2017 to 2019), as well as *ChocoMad* in Madrid, *Alimentaria* in Barcelona, the *Salon du Chocolat* in Paris, *The Chocolate Show* in London, and even events in Dubai! Today, my chocolate is available in many countries, like Canada, the U.K., France, Spain, and as far as Japan. My next big goal is to make a big impression at the upcoming chocolate show in New York City.

I currently produce about 1.5 tons of chocolate annually, including 200 kg of couverture for specialty chefs who value my chocolate's flavors.

## What are your main challenges?

**Of course, nothing** is perfect, but I face no insurmountable problems. The key is quality sourcing. Exceptional cacao beans are the foundation of aromatic chocolate, so I pay close attention to origin and supply. My Ecuadorian roots and cultural knowledge give me an advantage in navigating local practices, which helps me avoid both quality issues and financial pitfalls in a country with loose regulations.

## How do you see your activity in 5 years?

**This is difficult** to say because I never followed a master plan. When I first traveled to New York, I never imagined living there. Now, my husband and I are quite satisfied with the business as it is. I understand there is always an expectation of growth, but for us, we are not chasing volume but prioritize quality and sustainability, both in the chocolate and in our lifestyle. Our current framework functions for both our chocolate and for our quality of life. Whatever steps come next, they will be focused on Ecuador, the farmers, and above all, customer satisfaction.

## What is a good chocolate?

**For me, a** good chocolate is a total experience. Of course, the color, the texture, the aromas, and all the technical criteria must be right, but the story of how it came to be matters just as much and is part of the experience and of the appreciation. I immerse myself and spare no effort in learning about the people and the places behind my beans. Many Bean to Bar makers exist in the U.S., but my chocolate has an added dimension: it is crafted in Ecuador by Ecuadorians.

TANIA MOLINA WITH HER CHOCOLATE

## How do you see the cacao/ chocolate business in 10 years?

**I believe the** industry will place greater emphasis on quality and aromatic diversity. Consumers will eat better chocolate, not necessarily more of it. That said, U.S. chocolate consumption is relatively low if you compare it with countries like Germany, the U.K., or Switzerland, so there is still room for growth potential. But I am not a market forecaster, so I personally prefer to spend time perfecting my current product rather than trying to read the tea leaves.

# NICARAGUA

## ARGENCOVE CHOCOLATE
## - SALLY MCFADYEN

## CHOCOLATE MAKER INTERVIEW

| | |
|---|---|
| **Location** | NICARAGUA - GRANADA REGION |
| **Maker** | ARGENCOVE CHOCOLATE |
| **Interviewee** | SALLY MCFADYEN |
| **Comments** | REALIZED ONLINE, FROM ARGENCOVE'S FACTORY NEAR THE CITY OF GRANADA, BY LAKE COCIBOLCA |

## OVERVIEW OF NICARAGUA CACAO / CHOCOLATE SCENE BY THE AUTHOR

**Like all Mesoamerican** countries, Nicaragua has a long history with cacao. Although the Theobroma cacao fruit originates from the upper Amazon region, it has been endemic in the Central American isthmus since at least pre-Columbian times, and has been an important part of local culture for centuries. Nicaragua is rumored to be one of the countries where Christopher Columbus first tasted cacao. The fruit was initially cultivated in the central highlands and on the Pacific side, but nowadays the current expansion of cacao farming is taking place in almost every region of the country—if only to a small extent in many places.

Nicaragua's nutrient-rich soil, favorable climate, and easy access to both the Pacific and Atlantic oceans create ideal conditions for cacao cultivation. Moreover, the wide range of aromatic fine aroma varieties present in the country's many *terroirs* ensures a very broad diversity of flavors and aromas. Among chocolate professionals, Nicaragua is known as the land of flavor diversity. The national harvest is approximately 10,000 tons (2024) and is expected to grow exponentially to reach 25,000 tons by 2027. Even if this goal is realized, cacao will remain a very minor agricultural product in Nicaragua compared to coffee, which reaches about 140,000 tons every year.

Probably influenced by its southern neighbor Costa Rica, Nicaragua has been promoting sustainable farming and cacao with high organoleptic qualities for a couple of decades. Despite the increasing threats of diseases and extreme weather events, Nicaragua's cacao growth is likely to continue. If cacao's reference price remains above US$7,000, this trend may persist for many years.

## INTERVIEW OF SALLY MCFADYEN, CO-FOUNDER OF ARGENCOVE CHOCOLATE

ARGENCOVE
FINE ARTISAN CHOCOLATE

### How did you come to chocolate?

I AM A professional environmentalist, and my husband is a scientist. In 2015, we were living in Lima (Peru) and he was studying for an executive MBA. As part of the program, he developed a business case for a premium food business of which Farm to Table chocolate was a key product. Around that time, the Bean to Bar movement in the U.S. was becoming visible, and we decided to plan for a cacao and chocolate business. We wanted to control the whole chocolate production chain—from Farm to Bar—and looked at many possible countries in the region. Nicaragua ticked a lot of boxes. The soil and climate are perfect for cacao farming. Although coffee, sugar, and beef are key agricultural exports of the country, there is really interesting cacao here, the people are friendly, and it is relatively close to our main market, the U.S. We moved to Nicaragua in 2016 to start cacao farming. In 2017, we planted our 11 hectares of cacao with a combination of seed and grafted clones. We divided the land into multiple smaller plots in which we experimented with different cacao varieties, planting densities, and pruning techniques, all within an agroforestry system. Regarding varieties, we have some from the Costa Rican research center CATIE, others from Trinidad, Ecuador, and various local Nicaraguan varieties. All are classified as fine aroma. We collected our first harvest in 2019 and also sometimes buy wet beans (cacao en baba) from other farmers.

### Where is the name "Argencove" coming from?

WE THOUGHT LONG and hard about the name and used external help in developing the right word. "Argent" is from the Spanish word for silver—something precious, of high value. That aligns with our chocolate, which is indeed a luxury item. "Cove" refers to the location of our cacao farm, which lies below the Mombacho volcano near Lake Cocibolca—in sort of a cove.

### How did you learn how to make chocolate?

FROM THE START, we've always adopted a scientific approach to the business. We spent a lot of time and effort learning, guided by subject matter experts. For example, we attended the online course from the *Ecole Chocolat*, read many books and scientific papers, watched Manoa videos and used the resources of the Chocolate Alchemy website. But fermentation being so crucially important, we brought in external help: specialists from Trinidad, like Doctor Sukha from the University of the West Indies in Port of Spain, and also Cacao Services. We refined our post-harvest protocols—fermentation and drying—through much trial and error!

SALLY MCFADYEN WITH PART OF HER PRODUCTION TEAM

## What is your present situation?

**We farm on** 11 hectares, and plan to add another 5 hectares in the short term. Our staff numbers about 25, with seasonal increases during harvest and production peaks. Argencove sells products locally, but the large majority of our stock is shipped by air to the U.S., where we sell online as well as through retail outlets. A small amount is sent to Europe. We produce exclusively chocolate bars, nibs, and cacao butter—no by-products like cacao-liquor or dried pulp and juices.

## What are your main issues?

**We face the** usual production challenges, but nothing overwhelming. Thanks to our location and climate, we've avoided major fungal diseases, like Monilia, found in most cacao plantations in the region. This is a small developing country with decent infrastructure. The main highways linking us to the capital and its airport are quite good, so logistical problems are not insurmountable. Our most important challenge relates to the market: how can we educate consumers to understand, appreciate, and value quality chocolate like ours? We spend significant effort demonstrating the aromatic difference of our bars, the importance of clean, unadulterated and healthy ingredients, and cacao's social role in Nicaragua. We want to persuade the public at large that spending $10 to $15 for an Argencove chocolate bar is not only economically justifiable, but a delicious and healthy sensory experience. In other words, we must find the appropriate wording in the right environment to overcome the perceptions of what chocolate is and its price point. Since the First World War, industrial chocolate makers have promoted sugar and other non-cacao ingredients—such as vanilla and milk—so that many consumers today equate chocolate with sweetness rather than the deep, complex flavors and aromatic exposure you experience when tasting an Argencove bar. In other words, the Farm to Table story and full transparency are essential. I know we share this challenge with many other Bean to Bar makers, but this is really what concerns me the most. The new U.S. import tariffs add another layer of difficulty, driving up costs and making our mission even more important—and more challenging.

## What is a good chocolate?

**For me, a** good chocolate must be a complete and unique journey. It must tell a story in every aspect—the packaging must relate to the origin and the flavors. The look of the bar itself should not be too heavy and, of course, flavors should be clear and obvious and evolve in sensations by phases. Not a single "perfume" impression but a variety of scents and notes, that can even be contradictory at times, like a sharp acid note to accentuate sweetness.

## How do you see your activity in 5 years?

**We are currently** focused on expanding U.S. online sales. In five years, we expect to have a much larger commercial presence in North America, possibly with a retail shop. We will also serve more Nicaraguan customers through our own shop here. We will also be looking at the U.K. market. On the farming side, we will have another 10 hectares in production and, as our trees will be fully mature by then and therefore with a better yield, our harvest will be much larger. We will expect our chocolate production to have grown significantly, maybe triple or more from today. If all this materializes, we may also source more cacao from other farmers outside our plantation. You never know, with the current price level, more farmers might be interested. Regarding our product range, we will expand it to include more diverse chocolate products, but because of logistical constraints in Nicaragua, we will likely not move into delicate pastry items, like bonbons and truffles.

ARGENCOVE PRODUCT LINE

## How do you see the industry in 5 to 10 years?

**I think there** is a consensus on the continued growth of cacao and chocolate worldwide. This will happen through a wide diversification of the offer and of the consumption habits. The unavoidable higher prices of cacao beans and of chocolate will strengthen the relationship between producers and consumers, and increase awareness of the global impact of cacao farming. We are hopeful that with more flavorful and healthier chocolate, eating habits will evolve into more conscious consumption—moments of enjoyment tied to social or personal rituals rather than just mass consumption. A trend similar to the way people savor fine whisky or other specialty foods like caviar or exceptional pastries.

# SURINAME

## TAN BUN SKRATI N.V -
## ELLEN & RUTGER LEM

## CHOCOLATE MAKER INTERVIEW

| | |
|---|---|
| Location | PARAMARIBO |
| Maker | TAN BUN SKRATI N.V. |
| Interviewee | ELLEN LEM LIGTERINGEN & RUTGER LEM |
| Comments | REALIZED OVER WHATSAPP |

## OVERVIEW OF SURINAME CACAO / CHOCOLATE SCENE BY THE AUTHOR

**Suriname is part** of the Guiana Shield in South America, and 90% of the country is covered by rainforests. The cacao tree is native to Suriname, where the climate is ideal for cacao farming. The nutrient-rich soils of the coastal regions contribute to a complex pod flavor—cacao being just one of many sensations, as is typical for the Trinitario variety. Before mass production began around 1686, cacao was already being processed by the region's original inhabitants. At its peak—around the turn of the 19th and 20th centuries—Suriname was home to two million cacao trees, producing close to 4,500 tons of cacao annually. This represented approximately 5% of the world's harvest at that time. Today, production is so small that it doesn't even register in global cacao statistics. How did this happen?

Cacao was likely first planted for commercial purposes by Jewish settlers who had been relocated from Brazil, during the period when Suriname was still under English rule (1652-1667). At that time, the economy relied heavily on plantations worked by enslaved people. Before European colonization, Suriname was inhabited by several Amerindian indigenous tribes. It is not known how they organized cacao farming—whether in agroforestry systems, small gardens, or some other form.

During slavery in the 18th and 19th centuries, plantations could be quite big and Suriname's main crops were sugar, timber, coffee, and cotton. While many plantations also cultivated cacao—some as a main crop and others as a complement—overall exports remained quite low.

The abolition of slavery in 1863 triggered the rise of smallholder farming. However, this intensive cultivation method contributed to multiple outbreaks of witches' broom disease (WBD). At the time, this fungus had not yet

FARMING CACAO IN SURINAME

been identified, so no known remedy existed. It was eventually diagnosed in Paramaribo by Gerold Stahel, a Swiss scientist, but by then, cacao plantations had already been devastated before he could develop a viable response. This agricultural disaster triggered the launch of a breeding program on the neighboring island of Trinidad, which had suffered a similar catastrophe over a century earlier. Between the World Wars, wild cacao trees in the Upper Coppename River caught the interest of geneticists, as they were believed to be the first truly "wild" cacao trees ever discovered. Unfortunately, they were highly susceptible to WBD and did not survive.

Starting in the mid-1930s, British geneticist Dr. Francis J. Pound, from London's Imperial College, began working at the St. Augustine's campus of what is now the University of the West Indies (UWI). There, he developed a range of hybrid clones labelled "ICS" (Imperial College Selection), followed by an assigned number. These clones were both aromatic and more resistant to WBD.

Suriname attempted to revive its cacao culture using these more resilient trees. However, by then, Côte d'Ivoire, Ghana, and other European colonies in West Africa had vastly expanded their cacao farming, flooding the market with large volumes of low-quality beans at unbeatable prices. Since major confectionery companies showed little interest in aromatic cacao, Suriname had no outlet for its harvest. Despite this, the cacao industry limped on for a few more years, until the last cacao plantation finally shut down in 1997.

# TAN BUN SKRATI

Today, Suriname's minute local cacao production—less than 10 tons annually—is primarily used to produce the *Skrati* bar. Similar to the *Tablea* in the Philippines or the Mexican *Champurrado*, *Skrati* is made by grinding non-fermented dried beans in a pestle or a hand mill until the mixture becomes a semi-liquid cacao mass. Once cooled and solidified, it is grated and melted in a hot drink, often flavored with spices, milk, and, in Suriname, a local spice called bay rum, (known as "Beirum" ) which gives the *Skrati* its unique taste.

## INTERVIEW OF ELLEN & RUTGER, FOUNDERS AND CEOS OF TAN BUN SKRATI N.V

ELLEN & RUTGER

### How did you come to chocolate?

**Ellen** - I was born near The Hague in the Netherlands, and my mother was born in Suriname. I first came to Suriname as a seven-year-old on vacation and saw a cacao tree in my uncle's garden. As a child, I was puzzled: how could chocolate be made from that fruit? I was sure there had to be a faucet somewhere on the tree where you could tap out chocolate, which I imagined was flowing inside. Since then, I have always wanted to live there. Starting in 1996, I spent several months in Suriname every year, and in 2004 I settled there permanently. Rutger was born in the Netherlands and arrived much later, in 2011.

The director of the Women in Business Group (WBG) in Paramaribo, where I was working as a graphic designer—noticed my fascination with cacao and encouraged me to share it publicly during a national fair called "Made in Suriname" in 2009. Since the 1950s, imported chocolate in Suriname started replacing the locally produced *Skrati* in people's diets. Being able to afford foreign chocolate from the supermarket was considered more prestigious than making your own or buying local. However, my entry into chocolate making was driven by curiousity, instead of the patriotic "Made in Suriname" spirit. I simply wanted to finally solve the "cacao tree—chocolate" mystery that I couldn't understand as a kid, when observing the tree and its fruit in my uncle's garden. By participating in the fair, presenting and selling my products as a business (rather than as part of an art exhibition), and distributing them to friends and family, I realized that this could become a real business.

A year later, I bought my first grinder—a Premier mélangeur—and started teaching myself to make chocolate using online resources and multiple books. But the fermentation process proved a tough nut to crack. The results were widely inconsistent. When the Cocoa Research Centre (CRC) at the University of the West Indies (UWI) in Trinidad was unresponsive to my emails, I simply boarded a plane for Port of Spain to meet with the head of the department, Dr. Path Umaharan.

There, they explained the differences in fermentation outcomes, helped me achieve more consistent results, and gave me a tour of their facilities—in particular the WBD lab where they were growing the witches' broom fungus, a small chocolate-making and equipment set-up, and more. I learned to identify black pod disease by sight and smell. Their feedback helped me to refine my processes. Since that visit, the CRC has helpfully continued to support me online. In 2022, I eventually met Dr. Darin Sukha, who works in multiple areas along the cocoa value chain, especially focused on quality, flavor, and post-harvest treatment.

Simultaneously, back in Suriname, we were able to launch and grow our business thanks to the Muler family, who began replanting their 1.8-hectare family plot in the Saramacca district. Decades earlier, the four Muler brothers had planted cacao trees for their mother to make *Skrati*. From these original trees—some more than 60 years old —we've planted about 300 seedlings over the years.

As the plot is a mix of open field and agroforestry, we noticed that during the three consecutive La Niña years followed by El Niño, the 300 trees in the agroforestry and open-field sections fared much better than the others. With these observations in mind, we created a "cacao educational garden" as part of the LOKEO project based on developing agricultural and broader economic activities in the communal forests of Klaaskreek, a Maroon village in the Brokopondo district. This "forest garden" is a tool to promote agroforestry systems as a sustainable alternative to slash-and-burn open field agriculture in cacao farming.

For post-harvest, we ferment in food-grade plastic boxes that we heat-insulate in order to maintain an adequate temperature slightly around 40 Celsius. We do not use a tempering machine—instead, we temper our chocolate manually with silk cocoa butter.

CACAO WORKER FROM MAROON PLANTATION

## What is your present situation?

**We currently produce** about 200 fine-flavor chocolate bars every month, which we sell locally and export to chocolate boutiques in the Netherlands. We make batches with 72%, 80% and 100% cacao content, using white sugar cane which needs to be imported. We also produce and market various semi-finished cacao products and sell ingredients such as cacao nibs to be added to food and cacao tea made from the shells. From beans that we consider not good enough for fine chocolate, we press the fat and turn it into cacao butter lip balm sticks. From the cacao pod husks we extract the inner skin that, when soaked in water, produces an antiseptic, saponifying gel used as an ingredient in natural cosmetics and for haircare. Nothing goes to waste. Of course, we also sell the local *Skrati* which is used to make drinking chocolate.

We round out our business with dedicated workshops that share the story of Suriname's cacao. Participants learn how to make their own *Skrati*, dive into the aromas of fine flavor chocolate during a blind-tasting and even plant a cacao seed to take home and grow into a tree. We do not have a retail shop, so we distribute our products through specialty food stores, in some drugstores, and a souvenir shop too—both in Suriname and the Netherlands.

## What are your main challenges?

**As a general** rule, it is hard to do business in Suriname. While the country is part of the CARICOM common market, it is not included in the U.S. Caribbean Basin Initiative, which would grant tax-free access to the American market. While we can pay by credit card, we cannot receive payments through online gateways like PayPal, Zelle, or Stripe, as Suriname's banking system is not compatible with international online payment platforms. Most equipment and materials are imported, which makes it expensive to obtain basic equipment and accessories like packaging.

TRAINING IS A CRUCIAL TASK FOR US

Additionally, our practical concerns include improving indoor post-harvesting processes of fermentation and drying, especially managing acidic fumes during these stages, and the logistics of transporting chocolate in hot and humid weather. We developed our own indoor drying method out of necessity, as in Suriname the harvest of cacao is in the rainy seasons. It is simply not possible to dry beans when the relative humidity hits 100%. Our custom drying system allows us to bring out the full flavor potential hidden beneath the bitterness of freshly harvested beans. Subtle notes of fruits, spices, dairy flowers, and nuts are clearly identifiable in our chocolate. Thanks to the deep analysis and fermentation expertise of Elias Phiniotis—a Cyprus-born Canadian oenologist and winemaker—we were also able to develop a stable vinegar made from cacao pulp. In Suriname, this vinegar has long been used as a traditional remedy when mixed with coconut oil to soothe painful arthritic joints. Reviving such knowledge aligns with the philosophy of honoring both the science and the heritage of cacao.

## How do you see your activity in 5 years?

WE WANT TO continue growing through cacao-based agroforestry systems, while becoming fully sustainable in every aspect of our operation. This includes training and employing staff who can manage the business independently, creating a model that is not only scalable but also community-driven. A major goal is to deepen our collaboration with Maroon farmers, particularly those living in more remote regions of Suriname. These growers are key to reviving traditional, diverse farming systems, and we hope to support them through training, knowledge-sharing, and fair partnerships. For us, true sustainability means using 100% of the pod. We want to expand our cosmetics line, building on our work with cacao butter, cacao wine vinegar, and other pod components. There's untapped potential in these byproducts—not just as commodities, but as healing, value-added materials rooted in cultural tradition and natural science.

## What is a good chocolate?

A GOOD CHOCOLATE is one that does not hide anything. I believe that when the fermentation process and drying are done right, there's no need to use the roasting stage to mask flaws in the beans. If the cacao is of good aromatic quality, it will produce a chocolate with a complex flavor profile that retains the character of the fruit it came from and its origin, its terroir, and the hands that shaped it—from tree to bar. This is a good chocolate for me.

## How do you see the cacao/ chocolate business in 10 years?

IT'S HARD TO make long-term predictions, but here's an attempt to paint a possible vista of the future:

The cacao cultivation industry is facing a major crisis—with renewed large-scale outbreaks of fungal diseases, crop failures due to climate change, and continued pressures from industrial monocultures. This could spark a global rethinking about how we grow, value and consume cacao and its fruit. As mass chocolate production struggles with diminishing crops and resorts to alternative components, there may be a rediscovery of fine-flavor cacao—one that values quality, biodiversity, and cultural context over volume. Chocolate made from real cacao could once again become the niche product it used to be for centuries: rare, respected, and reflective of the land it comes from. Perhaps this will be cacao's "Reconquista"—a reclaiming of its place outside the candy aisle, away from anonymity.

As we say in Suriname: "Wan ogri kan tyari wan bun" ("Something bad can bring about something good").

Let's hope it does.

# CAMBODIA

## WAT CHOCOLATE - CORINNE POUILLIE & GAETAN BROSSEAU

## CHOCOLATE MAKER INTERVIEW

| | |
|---|---|
| **Location** | SIEM REAP - CAMBODIA |
| **Maker** | WAT CHOCOLATE |
| **Interviewee** | CORINNE POUILLIE & GAETAN BROSSEAU |
| **Comments** | INTERVIEW IN WAT CHOCOLATE FACTORY & SHOP IN SIEM REAP |

THE FACTORY IN SIEM REAP

## OVERVIEW OF CAMBODIA CACAO / CHOCOLATE SCENE BY THE AUTHOR

**Unlike its larger** neighbor Vietnam, until recently, Cambodia has never cultivated cacao, not even during the French colonial period. At the beginning of the 2000s, there were certainly attempts in the eastern and western regions of the country, near Vietnam and Thailand. However, in the absence of a political will and above all without a supporting ecosystem or local expertise, these initiatives failed to produce a sustainable activity. The few live cacao trees that remain from these projects have been abandoned.

In 2013, a Dutch chocolate enthusiast, Stefan Struck, partnered with Cambodian businessman Chanthol Chean, to launch an agroforestry cocoa plantation and a post-harvest center for fermentation and drying. They identified the Mondulkiri province, in the northeast along the Vietnamese border, as the most suitable region for cacao cultivation. The climate there is unusual compared to traditional cacao regions: hotter and drier, with an increasingly intense rainy season. Moreover, the province's famous red soil has often been depleted by monoculture.

It is therefore remarkable that Struck's company, Kamkav Farm, had by mid-2024 brought together more than 50 farmers—most of whom had moved specially to exploit cacao in the region. Together they now produce nearly 3 tons of dried, fermented beans from around 100,000 cacao trees. These figures may appear modest, but they mark the foundation of the first serious attempt to build a cacao ecosystem in Cambodia, with the potential to support "choco-tourism" across the province—part of a more global approach to promoting Cambodian cacao. Despite climatic challenges, particularly those linked to El Niño, the project continues to attract new participants, such as Wat Chocolate and Chocolate by H, both interviewed below.

## INTERVIEW OF CORINNE POUILLIE, FOUNDER AND CEO OF WAT CHOCOLATE

### How did you come to cocoa/chocolate?

MY BACKGROUND IS in costume design, while my husband Gaetan's is in hotel and catering management. Together we opened a restaurant in Paris, Le Canaillou. In 2012, we moved to Pondicherry, India, where we ran a guesthouse and organized motorcycle expeditions across the Indian subcontinent. There, we met Fabien and Jane Mason, founders of the Mason & Co. chocolate factory in nearby Auroville. Fabien shared his passion for cacao with Gaetan, who gradually trained as a chocolatier. Chocolate's journey from Bean to Bar requires many steps that make it a complex art to master.

THE FOUNDING TEAM

As soon as we arrived in Siem Reap in 2018, we began making chocolate from beans. The first challenge in spring 2019 was finding quality cacao beans in a country with no cultivation history. Fortunately, we met Stefan Struck, who was then pioneering cacao farming in Mondulkiri. By December 2019, we produced our first bars from a 90 kg batch of beans, and our company became one of the first major customers of Struck's company, Kamkav Farm.

In 2020, during the COVID pandemic, we used the downtime to do a lot of work in order to be ready for the restart of tourism in Siem Reap, which took place in 2023. Since then, our business has grown considerably.

### Where does the name "Wat Chocolate" come from?

WE NAMED OUR company Wat Chocolate to create a strong local identity that ties into what people associate with Cambodia—the word "*wat*" in Khmer means temple or pagoda. Given Cambodia's rich history of temples, especially in the Angkor Wat temple region on the outskirts of Siem Riep, using *Wat* as a business name evokes a sense of this heritage and signals that our chocolate is rooted in local culture. Our packaging also includes the words "Temple of Cacao" and we call our Phnom Penh shop "Temple Cocoa of Cambodia".

### What is your current situation ?

WE NOW OPERATE a well-located facility in Siem Riep, with quality equipment to manage each step of the Bean to Bar process. Our tools include a specialized oven for roasting, a winnower/cracker to produce the nibs, several stone grinding mills (mélangeurs) ranging from 4 to 18 kg that also "conch" (mix and soften) the chocolate, and a tempering machine as a possible complement to our silk butter tempering method. We also have a press to extract our own cocoa butter, and we make the silk butter ourselves for tempering. This approach also allows us to produce our own chocolate powder and to control the quality of the cocoa butter that we sometimes add to the recipe for our bars and, of course, to our white chocolate.

In terms of volume, we manufacture around 200 tablets a day, though our facility can support larger volumes. In spring 2023, we opened a new shop in Siem

Riep with a larger production lab and a staff of 12. This expansion allowed us to diversify: we now offer flavored bars at four cacao percentages (55%, 70%, 85%, 100%), along with candies, ganaches, and both plain and sweetened nibs. The increase in our production capacity also allows us to supply couverture chocolate to hotels and restaurants, especially through our second store in Phnom Penh, staffed by three employees. With more space, we have also started offering factory tours, which are highly popular with tourists. Gaetan and I continue to run the business, focusing exclusively on the Cambodian market, which is growing steadily thanks to both locals and international visitors.

A WIDE RANGE OF AROMATIC AND FLAVORED CHOCOLATE

A DEDICATED PRODUCTION TEAM

## What you main challenges today?

**OUR BIGGEST RECURRING** challenge is packaging. Processing high-quality beans—properly fermented and reasonably priced—is a constant concern, but packaging is particularly problematic. Having access to packaging consistent with our image and needs is made complicated here in Cambodia by complex import processes and the enormous volumes often required by producers in China or elsewhere. Since there are no specialized packaging producers in the country, we design our own and outsource production internationally.

## How do you see yourself in 5 years ?

**WE ARE FIRST** and foremost a profit-making business. We hope that Wat Chocolate will become Cambodia's leading chocolate company, fully vertically integrated. To guarantee a consistent supply of high-quality Cambodian beans, we plan to cultivate cacao ourselves. We are currently looking for investors and seeking land in the Mondulkiri province. Of course, cacao trees require three to four years before yielding harvests, but with Cambodia's 17 million people and fast economic growth, we see strong market potential.

## What is good chocolate for you?

**IN TERMS OF** taste, I am an atypical chocolatier—I actually like sugar! For me, a good chocolate should have a sweet taste. That's why I enjoy milk chocolate, and even white chocolate *(Author's note: white chocolate is not considered by experts to be "real" chocolate, as it has no cacao solids—just 33% cocoa butter, 33% milk powder, and 33% sugar)*!

# CAMBODIA

KATRIBU - CHOCOLATE

BY H - HANNAH LYN

## CHOCOLATE MAKER INTERVIEW

| | |
|---|---|
| LOCATION | PHNOM PHEN - CAMBODIA |
| MAKER | CHOCOLATE BY H |
| INTERVIEWEE | HANNAH LYN |
| COMMENTS | INTERVIEW AT THE CHOCOLATE BY H LAB IN PHNOM PHEN |

## INTERVIEW OF HANNAH LYN, FOUNDER AND CEO OF CHOCOLATE BY H

OUR LOGO

### How did you come to chocolate?

**BOTH MY HUSBAND,** Martin, and I are originally from Luzon in the Philippines. We have been living in Cambodia for over 30 years. I trained as a nutritionist and previously worked for the "Goya" food company in the Philippines.

While travelling in the U.S. in 2007, a visit to a dessert shop in Boston sparked my interest in making chocolates. I began with truffles and bonbons. To improve my skills and creations, in 2012 I completed the online course at *École Chocolat* and started building a chocolate laboratory at home. I produced a variety of ganaches and chocolate sweets, experimenting a lot with flavors. For example I made a ganache with the strongly aromatic Asian fruit, durian.

To further improve the quality of my production, in 2015 I travelled to Belgium, near Ghent, to attend the *Master Chocolate* course offered by Callebaut. There I deepened my understanding of chocolate, learning about the impact of the quality of the beans on chocolate's texture and aroma delivery, the diversity of the multiple variety of cacao beans, and the entire process of transforming beans into chocolate. Inspired, in 2016 I moved into the Bean to Bar business, while continuing to elaborate chocolate biscuits and other chocolate creations.

By 2021, I was gradually expanding my small-scale production when the pandemic hit, temporarily stalling progress. In 2023, I resumed field visits and purchased my first Premier stone grinder (mélangeur). To differentiate my products and respond to demand from diabetic customers, I created a sugar-free chocolate sweetened with stevia. Comparing it to Callebaut's 54% couverture, I found it even more enjoyable than the Belgian benchmark.

HANNAH & MARTIN IN THEIR CHOCOLATE LAB

## Where is the name "Katribu Chocolates" coming from ?

Our brand "Katribu" is named to evoke inclusivity. "Ka" comes from the Philippine language Tagalog (meaning part/membership) plus "tribu" for tribe. It's meant to celebrate the whole chocolate-process, especially including the farmers. This is because I recognize that what I do—chocolate making—is just one step in the whole cacao life journey. Without the farmers and everyone involved at every step of the value chain, there is no chocolate. Together, we form a tribe.

## What is your present situation?

**TODAY, WE ARE** still small. In my chocolate lab, I produce both chocolate confections and chocolate bars, which I sell via small shop retailers and directly to consumers, delivered to their homes by local "tuk-tuk" (motorized three-wheelers) drivers. Martin contributes mainly on the technical side, managing the machines. Sales vary month to month, but this year I expect to sell hundreds of bars and confections.

HANNAH LYN

## What are your main challenges?

**BEING SMALL AND** located in Cambodia presents many challenges. First, sourcing beans is costly. I buy some beans from *Kamkav* in Mondulkiri, the northern region of the country. But their harvest remains very small, and the quality is uneven. I have reservations about the consistency of their fermentation process, which sometimes results in bitter aromas. Buying from Vietnam is another option, but in small quantities, it is hard to compete with big names like "Piratos" of Holland. Having roots in the Philippines, I also buy beans and sometimes nibs from there, but shipping costs are higher. Sugar is expensive at about US$1.80 per kg. Although some cane is grown locally, Cambodia imports most sugar from Thailand. Finally, packaging is a headache because my small volumes prevent me from buying from China, where they require large minimum quantities or quotas.

## How do you see your activity in 5 years?

IN ABOUT 5 years, I hope to transfer chocolate-making skills to Cambodians, potentially through a cooking school for girls—perhaps partnering with the "Don Bosco" school network in Cambodia, that provides vocational education and formation of life skills for disadvantaged and out-of-school youth. That could be a good match! Afterwards, we would move back to Luzon to establish a chocolate laboratory back in the Philippines.

ROOM TO GROW

## What is a good chocolate?

A GOOD CHOCOLATE must have an attractive appearance—shiny and consistent—and a compelling story that convinces me it is farmed in a sustainable way, both ecologically and socially. Then, on tasting, it should release identifiable flavors that gradually evolve into a multitude of aromatic sensations. This is a "total experience", and to be honest, such chocolate is rare.

## How do you see the cacao/chocolate business in Cambodia in 5 years?

CACAO AND CHOCOLATE in Cambodia are poised for growth. Although the country is small, with 17 million inhabitants, there is significant potential. I see increasing appreciation for "real" chocolate, distinct from the overly sweet industrial products widely advertised here. Over time, more farmers will become involved in cacao cultivation, and Cambodian chocolate makers might eventually source 100% of their beans locally—perhaps in 10 years, if not 5.

# MALAYSIA

## CONCIERGE CHOCOLATE - NING-GENG ONG

## CHOCOLATE MAKER INTERVIEW

| | |
|---|---|
| LOCATION | KUALA LUMPUR |
| MAKER | CONCIERGE CHOCOLATE |
| INTERVIEWEE | NING-GENG ONG |
| COMMENTS | REALIZED AT CONCIERGE CHOCOLATE'S PRODUCTION SITE IN KUALA LUMPUR (MALAYSIA) |

## OVERVIEW OF MALAYSIA CACAO / CHOCOLATE SCENE BY THE AUTHOR

**Both the Malaysian** peninsula and the island of Borneo are ideal for cacao cultivation. The combination of a hot, humid climate and nutrient-rich soils creates near-perfect conditions for cacao farming. Cacao was first sighted in Malaysia at the end of the 18th century, during the Dutch's control of the port of Malacca in 1778, but the cacao trade only began in earnest in 1853 with the establishment of larger plantations. Under British rule, cacao cultivation spread widely across the country, especially in Sabah and Sarawak on Borneo. British companies established grinding factories and began producing chocolate for both the local market and export.

After independence in 1957, Malaysia expanded cacao production further to meet Europe's growing demand. In the 1970s, large tracks of farmland were opened up, and the high cacao prices at the time encouraged the use of high-yielding varieties to meet the sustained demand. At its peak, Malaysia's annual cacao harvest reached 247,000 tons. The country also began processing beans into cacao butter and powder, as well as making cacao by-products (cacao liquor) for the chocolate industry abroad. Average yields reached about 900 kg per hectare—nearly double the global average of less than 500 kg per hectare.However, the arrival of an insidious cacao pod borer (CBP) gradually reduced yields. At the same time, global cacao prices fell due to rising production in West Africa. As a result, many corporate farms switched to oil palm, a crop requiring far less labor. Palm oil demand then began its exponential, long-lasting rise.

Today, the Malaysia Cocoa Board oversees a harvest of less than 1,000 tons annually. Yet thanks to bean imports, the country processes about 240,000 tons into cacao butter, powder, and chocolate through ten factories. Malaysia now has the second-largest cocoa processing industry in Asia/Oceania and ranks fifth worldwide in grinding capacity, after the Netherlands, the U.S., Côte d'Ivoire, and Germany. Despite this, domestic chocolate consumption remains low. Malaysians only eat about 500 grams of chocolate per person per year, compared to nearly 11 kilograms per person in Switzerland or the U.K. Competition from imported brands is high. Given Malaysia's excellent natural conditions for cacao cultivation, the seemingly ever-growing global demand, and the recent rise in cacao prices, the country may well revive its farming sector while simultaneously developing a reputation for fine-flavor cacao.

## INTERVIEW OF NING-GENG ONG, FOUNDER AND CEO OF OF CONCIERGE CHOCOLATE

### How did you come to chocolate?

Before cacao, I worked as a developer of computer applications for various industries, from education to financial services, and including publishing. For as long as I can remember, I have always been attracted and even fascinated by the wide scope of flavors emanating from fermented foods. Since cacao beans undergo a fermentation process immediately after harvest, it felt natural for me to turn that passion toward chocolate. Being born in Malaysia, I also was curious about Malaysian cacao. Discovering that my country not only has ideal conditions for growing aromatic cacao varieties but also used to be a cacao powerhouse only deepened my fascination, increasing further my thirst for learning—and eventually made me take the plunge.

NING-GENG ONG

I treated every problem or barrier as an opportunity. For example, when I first contacted the Malaysian cacao board to ask where I could buy beans from indigenous farmers, I was told that they had no such directory but were organizing a cacao conference in which indigenous communities would participate. I immediately went, spoke with about 50 local small farmers, and then traveled to their villages to learn firsthand how they plant, grow, and care for cacao trees. This experience was fascinating and a real eye-opener for me. It is also why I now apply the agroforestry model I discovered there on my own plantation, with a 1:1 planting ratio (planting 1 cacao tree for each other tree species).

I came back from that experience totally energized and began making chocolate in my kitchen. Like most Bean to Bar beginners, I experimented with roasting cacao in my oven, grinding with a mortar and pestle, and mixing with a food processor! The resulting chocolate was definitely not up to specialty level, but it had generated a myriad of aromas that confirmed my passion. I then launched my company Concierge Chocolate with the intention of letting more people experience Malaysian-origin chocolate—chocolates that are the closest possible to the fruit and its *terroir*.

## Where is the name "Concierge Chocolate" coming from?

I SEE MYSELF as a bridge between people and the fruit. For me, "terroir" means not only the soil and ecological conditions—shade, water, air, wind—but also all the external factors that affect the tree and its pods throughout their life. The name "Concierge" underlines my role of bringing this terroir-driven experience of cacao to people and suggests a service-oriented approach. My vision is not just selling chocolate, but guiding people through a curated chocolate experience: tasting, learning about origins, and discovering flavor nuances.

## What is your present situation?

I SOURCE EXCLUSIVELY from Malaysia, across multiple states like Kelantan, Johor, Sabah, Pahang, and even Kuala Lumpur, working with indigenous communities including the Semai, Temuan, Temiar, and Dusun. About three hundred families in these groups supply half of the beans I process. Notably, some of the Malaysian beans I use won two "Cacao of Excellence" awards at the Paris *Salon du Chocolat* in 2021. Each origin has its own aromatic identity and uniqueness. We collect the pods and process them at our purpose-built post-harvest center in Raub. As fermentation is one of the main revealers of cacao aromas, I am very meticulous about the processes. Contrary to many fermentation operations, my procedure is not time-driven. We move along the fermentation steps depending on the evolution of the temperature, the pH, the sugar content, the size of the beans, the drainage, the taste, the smell, etc. We sometimes add heat or cold to accelerate or slow-down the proceedings. Yes, we spend a lot of effort and attention to the fermentation stage. Afterwards, the beans are dried on bamboo tables in a greenhouse. To be able to express each origin's personality, I do not blend origins. This also improves the traceability and sustainability of my chocolate.

Today, Concierge Chocolate produces about six tons of chocolate per year, offering chocolate bars, bonbons, and couverture chocolate for chefs and hotels. I enjoy researching and applying or tweaking traditional recipes to create new confections. Flavor experimentation is a big part of our identity. Creating new aromatic combinations is one of my favorite activities. We have a retail presence in two of Kuala Lumpur's best shopping centers and sell through baristas and, of course, online as well.

## What are your main challenges?

**First, we're very** happy to have survived the pandemic, which was a real challenge in itself. Now, the tasks that require most of my time and my efforts are human resources and bean sourcing. Over the years, Malaysia has lost most of its cacao farmers to palm oil or to rubber, leaving only a very small pool of providers. I hope the current hike in cacao prices will motivate more Malaysian farmers to devote some of their land to cacao farming. In our country, we are lucky to be able to produce up to 3 or even 3.5 tons of cacao pods per hectare, which is 3 times the world average. But, I must add that managing staff, as we are now over 20, requires a lot of effort. I insist on having dedicated and well-trained staff in order to maintain the quality of our chocolate. Staff well-being is also important to me, and we employ people with disabilities across various sites.

## How do you see your activity in 5 years?

**My goal is** to bring the beauty and richness of Malaysian cacao to chocolate lovers worldwide, but how exactly and how fast this will happen, I cannot predict—there are many external factors beyond my control. What I can say is that in 5 years, I will still be producing exceptional chocolate with Malaysian cacao beans.

## What is a good chocolate?

**My definition of** a good chocolate is influenced by a wine sommelier course I once took. I believe there must be an intentional approach to express the *terroir* in the chocolate. I believe that chocolate makers do not create flavors; they inherit them from nature. My goal, therefore is to "stay out of nature's way." A good chocolate should be crafted for the sole purpose of exalting the naturally created aromas of the cacao fruit. It must be flavor-driven—never vanity-driven, never cost-driven—a pure quest to transmit nature's gift.

## How do you see the cacao/chocolate business in Malaysia in 5 years?

**Over the next** 5 years, I anticipate steady growth in Malaysia's cacao and chocolate scene, fueled by increasing demand for single-origin and ethically sourced products, such as those crafted by Concierge Chocolate. The market is expected to expand by more than 7% annually, with premium chocolate—our specialty—surging ahead even faster. Tastes are shifting: health-conscious consumers are embracing authentic, single-origin chocolate, and appreciating craftsmanship over mass production. Malaysia is carving out a name as an exciting origin for fine chocolate, strengthened by innovation in local flavors and commitment to sustainable practices. As I always say, flavor comes first!

# TAIWAN

FU WAN - WARREN & AUDREY HSU

## CHOCOLATE MAKER INTERVIEW

| | |
|---|---|
| Location | TAIWAN - DONGGANG TOWNSHIP |
| Maker | FU WAN |
| Interviewee | WARREN & AUDREY HSU |
| Comments | REALIZED AT FU-WAN'S HEADQUARTERS AND AT A FARMER'S PLANTATION |

FU WAN CHOCOLATE
福灣巧克力

THE COMPANY LOGO

ARTISTIC PACKAGING

## OVERVIEW OF TAIWAN CACAO / CHOCOLATE SCENE BY THE AUTHOR

**Located about one** hundred kilometers off the southeastern coast of China, Taiwan—historically referred to as the Republic of China (and currently claimed by the People's Republic of China)—has no longstanding cacao culture. Although Portuguese explorers named the island "Formosa," they never settled there. The Dutch and the Spaniards followed, establishing bases on the island during the 17th century. It is possible that cacao was introduced then, either through Spanish-supplied Mexican *Criollo* beans arriving from the Philippines or *Forastero* beans brought via Dutch Indonesia from Brazil. Both colonial powers left in 1662, when Koxinga's regime and later the Qing dynasty took control until Japan's invasion in 1895. By then, cacao cultivation—never truly established—had disappeared.

A new chapter began in the 1920s, when Taichiro Morinaga, owner of a Japanese candy company, planted vast areas in eastern Taiwan with Indonesian cacao to meet Japan's burgeoning demand for chocolate. This venture ended with World War II. No further development occurred during the Chinese Civil War of 1949 or the decades of martial law that followed.

During Taiwan's 1980s liberalization phase, several initiatives attempted to revive cacao farming, including efforts to resurrect the Morinaga fields. However, battered by successive typhoons, they had practically vanished. Taiwan's high labor costs further reduced cacao farming's attractiveness, and the short-lived cacao revival quickly faded. Interest only truly resurfaced in the early 2000s. One of the pioneers of this "new wave" was farmer Chiu Ming-Song in Pingtung County. Though his legacy is sometimes debated, he played a pivotal role: actively distributing seeds, fostering collaboration among local growers, and earning recognition as the grandfather of modern Taiwanese cacao. Today, roughly 300 farmers cultivate a total of 250-300 hectares. In 2020, the last estimated harvest for the whole island was less than 5,000 tons of cacao pods, or about 300 tons of dry beans. Consequently, most of the small but growing number of Taiwanese craft chocolate makers and Bean to Bar makers import beans from Papua New Guinea and other established origins.

Taiwanese cacao bean varieties are heterogeneous and remain largely unclassified, which hinders branding efforts to build a strong "Taiwan origin" identity. Yet, with demand for specialty and premium chocolates growing, and with global cacao prices surging in 2023, local beans are becoming more competitive against imported ones. This trend is encouraging local farmers to integrate cacao trees into their farms and is helping to strengthen Taiwan's entire cacao value chain.

FU WAN FOUNDERS AND MANAGERS, AUDREY AND WARREN, AND THE AUTHOR (CENTER)

## INTERVIEW OF AUDREY AND WARREN, FOUNDERS AND CEOs OF FU WAN CHOCOLATE

### How did you come to chocolate?

**WARREN: I TRAINED** as a culinary chef and ran a high-end restaurant, a catering business, and a boutique resort before I moved to chocolate. As a chef, I was always looking for interesting—meaning tasty—local ingredients to integrate into my food. I began experimenting with small quantities of chocolate using imported beans from Ecuador, Papua New Guiana and elsewhere.

One day, a local farmer came to me with cacao pods and asked me how he might turn them into chocolate—or at least generate income from them. Like many farmers in the region, he was primarily growing betel nuts. Chewing betel nuts is highly addictive, and the crop has been in decline due to its clear link to mouth cancer. Facing the decreasing demand for betel nuts, many growers were switching to avocados, coffee, or bananas. Cacao then presented an alternative, provided there were buyers ready to buy pods or beans at a sustainable price.

I have always liked helping farmers, and I started thinking about how to make it work. I figured the only way to pay them fairly was to add value to the fruit by processing it in such a way as to turn it into a high-value product. That is when I decided to make my chocolate using local cacao pods, and I came up with systems to use every part of the pod. Now I not only produce highly sophisticated aromatic chocolate bars and bonbons, but I also create and market multiple cacao by-products. For example, I make chocolate liquor (chocolate mass), a sparkling drink I call "cocoa-cola" from cacao syrup, and even sell the pods' shells to be used as natural "saucers." I also integrate chocolate into many of my restaurant dishes—one of my favorites is a chocolate curry. It is original and unexpected, and customers say it is really delicious.

## Where is the name "Fu Wan" coming from?

**Fu Wan Chocolate's** headquarters are on the bank of the Dapeng Bay, a national scenic area near the city of Kaohsiung at the southernmost tip of the Formosa island. Our location is renowned for its natural beauty and rich cultural heritage, and provides an ideal backdrop for our brand's philosophy of harmonizing with nature. In Mandarin, the "Fu" character means blessing, fortune, and happiness and the "Wan" character translates to bay. Together, "Fu Wan" can signify an auspicious or prosperous bay full of blessings and happiness, which reflects our mission to share joy, positivity and prosperity through our artisanal chocolates.

## How did you learn how to make chocolate?

**As a seasoned** chef, I knew I could understand and adapt the chocolate-making process, but I also felt I needed to learn more about the aromatic potential of the cacao fruit. So, both my wife Audrey and I attended professional chocolate-tasting courses from the ICCT (International Institute of Chocolate & Cacao Tasting). We completed Levels One and Two at the organization's headquarters in London, and Level Three, in Lima (Peru). This training prepared us so well that we have since been recognized with many international awards.

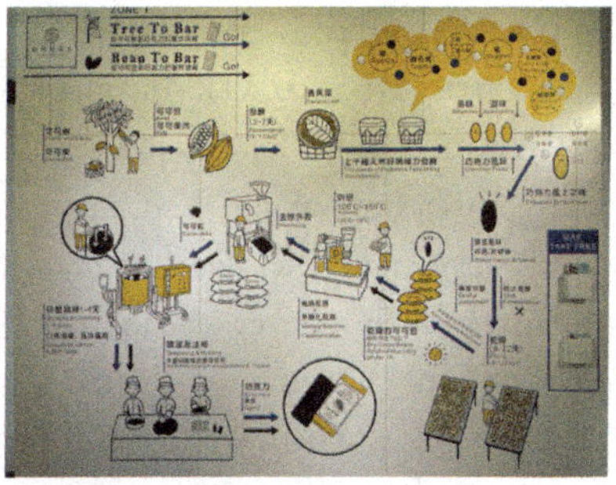


FU WAN'S METICULOUS PROCESS TO PRODUCE CHOCOLATE

In terms of spreading cacao skills to farmers, the regional government also invited a renowned professor from Taiwan's university to train farmers in the field, particularly in post-harvest practices and chocolate-making. Because Taiwan lacked a chocolate culture at the time, many participants had not even tasted chocolate. Still, the relative stability of cacao prices made the crop more attractive than betel nuts, whose market is highly volatile and lacks transparency.

At first, we bought bags of beans directly from farmers to make "single-origin" chocolate. But we soon realized that each batch—and sometimes even each bag—carried slightly different aromatic profiles, which made the chocolate's flavor inconsistent over time. That led me to start blending beans from multiple Taiwanese origins to create a more reliable flavor profile for customers. Eventually, I wanted to go further by shaping flavors myself. I began experimenting with differentiated fermentation protocols to see how I could introduce new tastes to chocolate. In 2017, I established my own post-harvest center, which gave me full control over quality from start to finish. and to enhance the beans' aromatic characteristics by developing new fermentation protocols. That's where I began developing new flavor fermentation methods—and over time, I've created around twenty original processes. One of them, I call "double fermentation," and as far as I know, it's unique. That is why I keep it as a trade secret! But the results speak for themselves—I now have a whole range of distinctively flavored chocolate bars that are highly appreciated by chocolate sybarites worldwide.

THE FERMENTATION BUCKETS, PREVIOUSLY USED FOR RICE

### What is your present situation?

**For the chocolate** business, we have a total staff of 20. That includes the fermentation center here at Fu Wan headquarters, our local shop, and the flagship store in Taipei. We exclusively source cacao from Taiwanese farmers. In Taiwan, the typhoon season falls in July and August, while winter—with temperatures dropping to 18 to 20°C under clear skies—runs from December to March. There are two main harvest periods: March to June and October to December, though the other months still yield enough beans for us to continue working. From cacao's flowering to a pod that's ready to harvest, it typically takes four to five months. We produce approximately ten tons of chocolate per year, making us the second-largest buyer of Taiwan-grown beans. And our products aren't just enjoyed locally—Fu Wan Chocolate is available in international markets as well, including Europe and the Americas.

### What are your main issues?

**Not everything is** perfect, but I do not feel overburdened by production issues. My main concern is really raising consumer awareness about premium chocolate and helping people understand why it commands its price. I spend a lot of time and effort to elevate the market. For example, the average Taiwanese person consumes about 1 kg of chocolate per year—mostly confectionery chocolate or chocolates bought as gifts. *(Author's note: to put that in perspective, in the U.K. and Switzerland it's around 11 kg per person, and in France, about 7 kg.)* So, my main issue is to educate the public about what "real" chocolate is—the aromas, flavors and characteristics that make great chocolate. My hope is that, over time, people will not only appreciate it more but also consume more of the premium chocolate that I produce.

## What is a good chocolate?

**Honestly, a good** chocolate is the one you love—the one you just can't stop eating. I am very open about personal preferences. I am especially interested in bars with local or interesting infusions, or a flavor I haven't tried yet. Of course, I have my own tastes in terms of flavor strengths and types; but this is me, it's not an objective standard.

## How do you see your activity in 5 years?

**In five years,** I hope to have shops abroad, perhaps in the U.K. and/or the U.S. Like here, the U.K. is an island nation, and I love the kind of open culture it has gained from other countries. People there are genuinely curious about new flavors. The U.S., of course, is also very appealing. It is a huge market, and it's where the Bean to Bar phenomenon really took off. Things move quickly there. I've also thought about launching a second brand—one focused entirely on the constantly evolving flavors I create. That's the part I love most about my work. Last year alone, I developed over forty new aromatic chocolates. I'm always experimenting with fermentation and exploring how fruity flavors—such as apple, grapefruit, passionfruit, and strawberry—can be mixed to incorporate them into the fermentation process so they can blend in beautifully with the chocolate.

I am also experimenting with storing cacao beans in vacuum-sealed bags, with the goal of aging them much like fine wine. Over time, I believe the beans will undergo subtle chemical and aromatic improvements—smoothing out harsh flavor notes. In the future, I plan to produce chocolate from beans aged more than 20 years, releasing them as special "vintage" editions. And of course, just as with fine wine, the rarity and uniqueness of these aged beans could increase their value over time.

## How do you see the cacao / chocolate business in Taiwan in 5 years?

**I think we're** seeing a lot of change right now in the cacao and chocolate world globally. But honestly, I can really only speak for what I see here in Asia. It's clear that consumption across the region and in Asia as a whole will keep growing. Culturally, chocolate in Taiwan is often associated with special occasions and gifts, rather than daily appetites. I believe that growth will come with new kinds of chocolate—products and flavors that connect with the Asian palate and culture. What's really interesting is how the Asian palate is beginning to shape its own chocolate identity. I think we'll see more "unconventional" combinations, things that might surprise people elsewhere but feel perfectly natural to us here. And that's great, because it shows how chocolate is becoming part of our own food identity.

FU WAN'S RANGE OF EXQUISITE CHOCOLATE BARS

# VIETNAM

## MAISON MAROU - VINCENT MOUROU

## CHOCOLATE MAKER INTERVIEW

| | |
|---|---|
| LOCATION | HANOI - VIETNAM |
| MAKER | MAISON MAROU |
| INTERVIEWEE | VINCENT MOUROU |
| COMMENTS | INTERVIEW FROM HANOI TO SAIGON VIA TEAMS |

## OVERVIEW OF VIETNAM CACAO / CHOCOLATE SCENE BY THE AUTHOR

WHILE VIETNAM IS now widely known for its coffee, cacao farming is slowly but surely becoming a significant part of its agriculture—even though the total cacao harvest is relatively small (around 5,000 tons). Cacao was actually introduced by the French at the end of the 19th century. It was mostly grown for local use, with a few beans sold around the region. The plantations were mainly in the south, near Saigon (now Ho Chi Minh City), where the tropical climate and soil are ideal for cacao. After independence from France in 1954, Vietnam restructured its economy and agriculture. Cacao was identified as a viable cash crop, alongside traditional Vietnamese ones like rice, coffee, or rubber. But since those other crops were already well established, they received most of the government's attention and funding, so cacao never really took off at that time.

Things began to change after the 1986 economic liberalization. The government started encouraging crop diversification and exports, and cacao once again became an opportunity. In the 1990s, major international chocolate companies like Cargill and Mars began looking for new sources of cacao to meet rising global demand. Vietnam, with its favorable climate, fertile land, and relatively low labor costs, became very attractive. That led to the creation of new cacao plantations in southern provinces such as Ba Ria–Vung Tau, Dong Nai, and Ben Tre. The government also stepped in to support the sector—offering technical assistance, access to credit, and promoting research to improve productivity, quality, and sustainability. Interestingly, Vietnam never joined the International Cocoa Organization (ICCO), but that hasn't stopped its progress.

Today, in the 2020s, Vietnam is emerging as a significant player in the cacao market. Both production and domestic chocolate consumption are growing—although the average Vietnamese still eats only about 200 grams of chocolate per year, compared with 4.5 kg in the U.S. or 11 kg in Switzerland. Vietnam now exports both raw beans and semi-finished cacao products to Europe and North America.

Despite challenges like plant disease, market price fluctuations, and competition from other producing countries, the outlook remains very positive. The quality of Vietnam's cacao—mostly Trinitario varieties—and the use of advanced farming and post-harvest techniques mean that Vietnamese beans often sell at a premium, above the prices set in New York or London. That translates into better farmgate prices for local farmers.And with the rise of specialty chocolate and the Bean to Bar movement, Vietnamese producers are also benefiting from growing consumer awareness of ethical, sustainable, and traceable sourcing. Given the current global situation—with prices climbing and yields dropping in many traditional cacao-growing regions—Vietnam is well positioned to expand its role in the global chocolate industry in the years to come.

## INTERVIEW OF VINCENT MOUROU, FOUNDER AND CEO OF MAISON MAROU

### How did you come to chocolate?

**I WAS BORN** in France but grew up in the U.S., where I graduated from the University of Michigan with a B.S. degree in 1994. Over the years, I tried my hand at various things. I was a filmmaker before traveling to Vietnam in 2010. That's where I met Samuel Maruta, who had been a banker and was also in search of a new business adventure in order to remain in Vietnam, which we both loved. We quickly became friends. When we learned about Vietnam's cacao, we started looking for some beans. Thanks to Google, we found some in Southern Vietnam, but wanted to discover more. So we jumped on our motorbike and rode around the country. Amusingly, we often drove past trees we thought were cacao, only to realize they were actually coffee! This tells you a lot how little we knew about cacao back then.

VINCENT MOUROU CHECKING CACAO ROASTING

We started by making very small 1 kg batches using an Indian-made grinder and beans from farmers across 4 or 5 provinces. This helped us compare terroirs and understand how the origin affected flavor. It was very exciting to have so many different flavors and chocolate sensations within a single country. It became obvious that the terroir would be the key to defining our chocolate's identity and differentiate it from other chocolate.

From the beginning, we've always been very supportive of cacao farmers—for both ethical and practical reasons. Without farmers, there is no chocolate. We still work with most of the same farmers we started with, except for one. Today, we have a team of ten people dedicated to sourcing Vietnamese cacao and collaborating with farmers and fermenters to ensure the best quality and fair prices, working with farmers and fermenters in order to have the best quality and to ensure farmers receive an honest price.

Sam and I learned to make chocolate thanks to a German professor working with the German Aid Agency in Vietnam. He was training farmers and kindly gave us a few masterclass-style lessons. Later, the American aid agency USAID launched programs to expand cacao farming and strengthen the cacao value chain—helping farmers select varieties, prepare land, prevent diseases, and graft seedlings. Indirectly, that also helped us, even though we financed and developed Marou Chocolate entirely on our own.

We soon realized that farmers could produce more than we could buy. So as early as 2013, we started to buy beans for export—first to the U.S., and later to Europe and other regions. Through this process, farmers began connecting with others from different regions and comparing their crops. That's how we started building Vietnam's first cacao *terroir* identity.

Unlike wine and other food products such as cheese in France, cacao does not have an AOC (Protected Designated Origin system). We're a long way away from that kind of system in cacao, if it is even possible. I think it would be hard to create a protected origin classification for cacao because the fermentation process is so impactful on the flavors that just the geographical origin of the fruit alone cannot fully define the *terroir* of the beans and therefore of the chocolate.

## Where is the name "Maison Marou" coming from?

"*MAISON*" IS OF course the French word for house, which in French branding often refers to haute couture "*maisons*" of high craftsmanship and *savoir-faire*—places curating an entire artistic culture around their high quality brand, as Maison Marou aims to do around Vietnamese cacao. Our shop format (incorporating an open kitchen where you can view the process) where we give classes, creates a kind of home for the vision of chocolate by our customers. As you have probably guessed, "Marou" is a contraction of my co-founder Sam's and my last names (Maruta and Morou).

## What is the present situation of "Maison Marou"?

WE'VE GROWN INTO quite a large company. With 16 retail locations—including flagship stores offering immersive chocolate-making experiences and cafés—we now have about 420 people, more than half of them working in retail. Marou's trajectory really changed when we opened our first retail store on Calmette Street in Saigon in 2016. Until then, we had a strong reputation but were still quite small. Around the same time, the Vietnamese middle class began embracing locally made, high-quality chocolate like ours, and our sales took off. We now export to twenty countries, and have launched a slew of new chocolate products—chocolate bars, power bars, spiced chocolate, chocolate spreads, flavored nibs, and more. Interestingly, 85% of our sales are now domestic, compared with 25% when we first started. In volume, exports haven't dropped, but local demand has exploded thanks to our cafés and specialty stores. The Vietnamese chocolate scene has also blossomed, with about fifteen Bean to Bar makers now active in the country.

MAROU'S MAIN SHOP IN VIETNAM'S CAPITAL

## What are your main challenges?

**Sourcing quality beans** is clearly an important issue for us. Even before the recent weather-driven shortages, we were always searching for quality beans and skilled fermenters to process them. Climate change affects all crops, and cacao is particularly sensitive, especially to diseases—it takes three to four years before new trees bear fruit, so you can't quickly offset poor harvests. Because of these constraints, we've stopped exporting beans—we now keep everything we can buy. To address this, we're deepening our work with farmers and promoting agroforestry. Any real increase in Vietnam's cacao production will only happen if we, all the stakeholders, collaborate closely.

## How do you see your activity in 5 years?

**As the retail** side of Marou keeps growing, I see us opening more immersive chocolate experience stores, both within Vietnam and abroad—first across Asia, in places like Singapore and Japan, and eventually in Europe. As this expansion will require us to scale up production, we will be extremely vigilant to retain our high level of quality and sustainability This is in fact, our greatest challenge: how to grow while staying true to ourselves.

## What is a good chocolate?

**For me, the** characteristics of a good chocolate could be summarized as "No off-flavors, not too astringent, and not too bitter." I prefer notes of fruitiness and scents of spices. I like chocolate that is bright and fresh—not bland.

My preference goes to fresh chocolate. Some makers now age their beans or chocolate for months, producing very smooth, mellow bars. Personally, I find those lack the liveliness, freshness and personality that I enjoy. We process beans within weeks of harvest and fermentation, and that freshness makes all the difference and gives the best results.

SEEKING PERFECTION

## How do you see the cacao/chocolate business in 10 years?

**I think that** chocolate can only go up from here. Sourcing limitations and issues will maintain the price of cacao beans well over US$6000 per ton. (Note from Author: prices fluctuate, after peaking at US$12,000, and have remained well above $8000.) At the same time, quality will continue to improve, thanks to better farming techniques and fermentation processes. Large industrial companies will use more cacao substitutes to maintain margins, which will widen the gap between candy and what I call real chocolate. Through better communication, consumers in traditional chocolate markets are gradually accepting the higher cost of real chocolate, while new markets are treating chocolate as a specialty food. So yes, I see a bright and promising future for cacao and chocolate.

MAROU'S RANGE OF CHOCOLATE BARS

# ZERO CARBON

**Sailing Toward a Sustainable Chocolate Future:**

ACROSS THE OCEANS, a quiet new trend is underway in the way chocolate reaches our tables. Specialty and Bean to Bar chocolate makers, many like the ones profiled in this book, are increasingly demanding sustainability at every step—from the cacao farms in the tropics to the factories and shops in Europe and the Americas. The Bean to Bar chocolate industry has long been at the forefront of environmental and social responsibility. Leading makers are deeply aware of their role in creating a sustainable cacao supply chain that respects not just ecological systems, but the communities that produce the beans. This includes sourcing cacao from farms that avoid child and slave labor, ensuring farmers are paid fairly, and requiring that cacao cultivation does not contribute to deforestation—standards now reinforced by E.U. regulations. At the production stage, many chocolate makers are transitioning to clean energy in their production sites—using green packaging, and adopting other eco-friendly practices to reduce their carbon footprint.

Yet sustainability in chocolate extends far beyond farms and workshops. The most visually striking potential is at sea, where carbon-free shipping is slowly and steadily becoming a trend. Since the early 2000s, specialized shipping companies have begun transporting cacao beans from their countries of origin in the tropics to chocolate-making countries in Europe and the U.S. using sailing ships. These ships—both restored vintage vessels and purpose-built modern craft—now carry cacao, chocolate, coffee, and other tropical goods across the Atlantic, proving that environmentally responsible transport can be both feasible and practical. The journey from cacao plantation to finished bar often spans thousands of miles, traditionally relying on fossil-fuel-powered ships that generate significant $CO_2$ emissions. In response, a new wave of innovators is hoping to transform cacao transport. Their initiatives blend traditional sailing techniques with modern technologies—advanced sails, electric motors, and satellite navigation—while also offering educational and immersive experiences for crews and passengers.

As described more fully below, some like the Dutch company "Fairtransport," use vintage restored schooners, while others, such as France's "Grain de Sail" have built state-of-the-art cargo boats propelled by the wind with modern sails. Meanwhile, several traditional shipping companies that still use conventional vessels now offer to offset their carbon emissions by purchasing carbon credits, which help fund reforestation and other renewable energy projects aimed at reducing atmospheric $CO_2$.

By combining carbon-free logistics, and innovative transport solutions, altruistic chocolate makers are not only starting to reduce their environmental footprint but also enriching the stories of their chocolates' origins. Each cacao bean or chocolate bar transported via these vessels represents not only an ethical choice but a good story as well. In a market increasingly conscious of climate impact, sustainable shipping can become part of the chocolate experience itself, enhancing flavor of the stories behind each bite.

Looking further ahead, several high-tech startups are developing ships powered by green hydrogen or green ammonia—fuels produced from renewable electricity that contain no carbon molecules and therefore emit no $CO_2$. Today, multiple companies operate zero- or low-carbon cargo ships across the oceans. Here are a few pioneers:

## Fairtransport:

**Sometimes described as** the initiator of carbon-free cargo shipping, this Dutch company began operations in 2008 from its base in Den Helder, just north of Amsterdam. Its stated mission and goal is to minimize the global $CO_2$ footprint, raise awareness of sustainable transport, and demonstrate that freight can be transported in an environmentally responsible way.

Animated by a strong sailor's mentality, the founders transformed a 1943 German mine sweeper wooden hull into an eye-catching 19th-century brigantine named *Tres Hombres*. The boat had to be registered in Vanuatu, the only country that authorizes commercial cargo ships to be 100% engineless. Crewed by a mix of professionals and volunteers, every winter the ship sails from Amsterdam to the Caribbean and South America to load cacao beans, rum, coffee, and spices for return delivery to Europe. Those transatlantic tours sound charming but are incredibly hard work. Export procedures from cacao countries require a heavy load of paperwork, and once at sea, the crew manages up to 17 sails without winches or synthetic ropes. Every time the boat changes course, they all need to be readjusted. Arrival and departure dates depend entirely on the wind, and the ship often anchors offshore, as maneuvering in port without tugboats is extremely difficult. Fairtransport's success has led to the addition of two more vintage ships focused on African and Mediterranean routes, with three additional vessels currently under restoration.

*TRES HOMBRES* MOORED

### Hawila:

**With the mission** of "transporting goods emission-free and inspiring action for sustainable change", this Danish cooperative is in the process of rebuilding an old two-masted Baltic sailing ship named *Hawila* to run entirely on wind power. It also serves as a platform for environmental education and sustainability awareness. Plans are underway to further enhance her energy efficiency by integrating wind turbines and solar panels, aiming to make her operations even more sustainable. Its maiden voyage is planned in 2026, in order to transport organic goods across the Atlantic between the Caribbean and Europe without generating emissions.

### Grayhound Ventures:

**From Sweden, this** company is using the U.K.-built *Grayhound* sailing ship, a replica of a three-masted British customs vessel, to offer "environmentally friendly adventures and cargo delivery, to the benefit of coastal communities worldwide." Launched in 2012, the ship serves dual purposes: as a sustainable cargo carrier and as a platform for adventure sailing experiences. Mostly operating in Europe, the team is hoping to develop larger sailing projects.

### EcoClipper:

**Founded in 2018,** this Dutch company is developing a fleet of emission-free, sail-powered cargo ships. In 2023, EcoClipper completed the retrofit of the 1912-built *De Tukker*, its first operational vessel, which successfully demonstrated the feasibility of sustainable maritime transport by carrying Dutch chocolate from Amsterdam to Porto with a small crew. Recovering from the near loss of its main ship, the company aims to resume regular carbon-free cargo transport from tropical countries to Europe in 2026.

Looking ahead, EcoClipper is designing the *EcoClipper500*, a modern clipper-inspired vessel capable of carrying up to 500 tonnes of cargo, 12 crew members, 36 trainees, and 12 passengers. The company's vision combines traditional sailing techniques with modern technology, expanding the fleet while offering hands-on sailing experiences for passengers who want to learn traditional maritime skills.

### Grain de Sail:

**This Brittany-based French** company is driven by a mission of "Taste, Adventure, and Sustainable Development." To bring this vision to life, Grain de Sail produces and transports coffee, chocolate, rum, and wine using two purpose-built modern sailing ships, *GDS I* and *GDS II*, with a third vessel set to launch in 2027. By carrying both their own products and customers' cargo, the company demonstrates how carbon-free shipping can be practical and profitable. Their chocolate is crafted from organic cocoa sourced mainly from small-scale farmers in the Dominican Republic and Peru, with above-market payments supporting the transition to sustainable farming. In Morlaix,

Brittany, Grain de Sail also operates a chocolate factory and roasting facility, where visitors can explore the production process, participate in workshops, and learn about chocolate firsthand.

**Brigantes:**

INITIALLY FOCUSED ON shipping and roasting coffee, the Austrian-Italian initiative Brigantes is now consolidating various sailing cargo ships used by European importers of tropical goods. In collaboration with other organizations, Brigantes has already transported cacao via sail cargo—for example, partnering with the ship *Avontuur* to move cacao from Nicaragua to Hamburg, marking the first docking of a sailing cargo ship in the port in 20 years. The company is in the last phases of restoring the historic sailing vessel *Brigantes*, originally built in 1911, in Trapani, Sicily. The brigantine will sail powered entirely by wind, transporting goods such as coffee, cacao, chocolate, and rum from the Caribbean and Central America to Europe by following traditional trade wind routes.

**SailCargo Inc.:**

THIS CANADIAN COMPANY believes that eliminating fossil fuels from the maritime sector is achievable by using advanced technology and simple techniques. The construction of their three-masted cargo schooner *Ceiba* built in Costa Rica was a collaborative effort involving shipwrights, carpenters, business professionals, and sailors, all dedicated to proving the value of clean shipping and to inspiring change in the shipping industry. While primarily relying on wind power and following customary trade wind routes between the Americas, *Ceiba* is equipped with auxiliary electric motors for maneuvering in ports and during low wind conditions. *Ceiba* is designed to transport up to 280 tons of cargo, including goods such as coffee, cacao, and chocolate.

## Zero emission transport sailing forward:

ALTHOUGH VINTAGE SAILING presents a charming and romantic experience, the future of zero emission transport is in new technologies. To give just a few examples, technologies such as ultra-efficient orientable wings to catch the wind, robotics to constantly position the ship into the best orientation, satellite navigation to take advantage of meteorological conditions, and solar panels to power harbor maneuvers, will eventually make carbon-free shipping reliable and economically viable. Many enterprising shipping companies worldwide are investing towards a carbon-free future, following specialty chocolate makers' goals to eliminate global warming contributions in their transportation as well as in manufacturing processes. Let's hope that these optimistic trends will grow and flourish in order to help fulfill the motto/prediction of the indigenous Maya cacao farmer Jaun Cho, founder of Ixcacao chocolate in Belize: "Cacao will save the planet!"

# EPILOGUE

This voyage through the world of tropical chocolate makers reveals a landscape in transformation. Across humid valleys and forested hills, new voices are emerging — people who have chosen to reinvent the way cacao and chocolate are grown, crafted, and understood.

Those who devote their energy to this "new chocolate scene" often come from paths far removed from traditional farming. They are agronomists, entrepreneurs, engineers, and idealists, united by curiosity and care. Their ambitions reach well beyond the bean itself: they seek to capture aromas, to nurture social well-being, and to align their work with the ideals of global sustainability.

The chocolates they produce carry the imprint of place — the soil, the rainfall, the language, the local palate. Each bar reflects a particular culture and its idiosyncrasies, offering flavors that may surprise even the most seasoned chocolate lovers.

In many cacao-growing regions, a small but growing number of people now choose to savor these locally made chocolates. Their enthusiasm proves that excellence can thrive close to home, even in difficult economic conditions.

The emergence of this new class of chocolate is already reshaping the global market. It challenges established norms and justifies higher prices that safeguard both quality and the livelihoods of those who make it possible.

Global cacao production continues to expand by roughly 4.5% each year, and is expected to surpass seven million tons in the early 2030s. Climate fluctuations will bring uncertainty and price volatility, yet the overall trend remains upward — with values likely to stay above seven thousand US dollars per ton.

Unpredictable though it may be, the horizon is luminous. For cacao farmers and chocolate lovers alike, the future holds the promise of deeper connections — between people, flavor, and the landscapes where chocolate begins.

# "FROM BEAN TO BAR", A CONCRETE VISION

# NOTES

Made in the USA
Monee, IL
02 November 2026

67382835R00046